Isotopes: Essential Chemistry
and Applications II

Special Publication No. 68

Isotopes: Essential Chemistry and Applications II

The lectures delivered at a Residential School organised by The Royal Society of Chemistry in conjunction with The University of Surrey

University of Surrey, January 6th–8th 1988

Edited by
J. R. Jones
University of Surrey

ROYAL
SOCIETY OF
CHEMISTRY

LHEM
Sep/al

British Library Cataloguing in Publication Data
Isotopes: essential chemistry and applications
II.
1. Isotopes
I. Jones, J.R. (John Richards)
541.3'88

ISBN 0-85186-746-4

Published by The Royal Society of Chemistry,
Burlington House, London W1V 0BN

Printed in Great Britain by
Whitstable Litho Printers Ltd., Whitstable, Kent

SD
4/18/89
XK

Introduction

In the nine years that have elapsed since the first review symposium[1] on Isotopes: Essential Chemistry and Applications was held at the University of Surrey several important developments have taken place. The need for scientists interested in the synthesis and applications of isotopically labelled compounds to meet on a regular basis has been recognised in that two international symposia have been held on this topic. In both cases the proceedings have been published.[2,3] A third symposium is scheduled for Innsbruck in July 1988. In addition an International Isotope Society has been formed with the expressed aim of promoting the applications of isotopically labelled compounds.

In the first review symposium sponsored by The Chemical Society (as it then was) the main focus of attention was on compounds labelled with the isotopes of hydrogen and carbon and the analytical methods used, prior to their application in biosynthesis, drug metabolism and reaction mechanisms studies. Here the emphasis is on short-lived radioisotopes (^{11}C, ^{13}N, ^{18}F and ^{125}F), other analytical methods, particularly radiochromatography and autoradiography, and developments in existing methods, particularly nuclear magnetic resonance spectroscopy, since 1979. The industrial uses of radioisotopes are emphasised as also are the applications of radiolabelled compounds in molecular biology; applications of compounds, mainly labelled with stable isotopes, in medicinal chemistry are covered. The uses of radiopharmaceuticals for diagnostic purposes as well as positron emission tomography (PET) investigations are the subject of a separate chapter. Taken together with the proceedings of the first review symposium the two volumes cover the synthesis of a wide range of isotopically labelled compounds, the analytical methods used and many important applications.

J.R. Jones
Guildford
June 1988

References

1. Isotopes: Essential Chemistry and Applications (eds. J.A. Elvidge and J.R. Jones), The Chemical Society, London, 1980, 400p.

2. Synthesis and Applications of Isotopically Labeled Compounds (eds. W.P. Duncan and A.B. Susan), Elsevier, Amsterdam,. 1983, 508p.

3. Synthesis and Applications of Isotopically Labeled Compounds, (ed. R.R. Muccino), Elsevier, Amersterdam, 1986, 557p.

Contents

Contributors

D.B. DAVIES	Department of Chemistry, Birkbeck College, London.
P.S.G. GOLDFARB	Department of Biochemistry, University of Surrey.
D. HALLIDAY	Clinical Research Centre, Medical Research Council, Harrow.
F.M. KASPERSEN	Organon International, Oss, The Netherlands.
K. KRISTENSEN	The Isotope-Pharmacy, The National Board of Health, Copenhagen.
D.J. LESTER	ICI Physics Radioisotope Services, Billingham, Cleveland.
W.J.S. LOCKLEY	Fisons (Pharmaceutical Division), Loughborough.
V.W. PIKE	MRC Cyclotron Unit, Hammersmith Hospital, London.
D. SILVER	Amersham International, Cardiff.
M.A. WILLIAMS	Department of Anatomy and Cell Biology, University of Sheffield.

1
Organic Synthesis with Short-lived Positron-emitting Radioisotopes

V.W. Pike

MRC CYCLOTRON UNIT, HAMMERSMITH HOSPITAL, DUCANE ROAD, LONDON, W12 0HS, UK

1. INTRODUCTION

Chemistry with short-lived positron-emitting radioisotopes of the non-metals, principally ^{11}C, ^{13}N and ^{18}F, has burgeoned over the last decade. This has been almost entirely because of the emergence of positron emission tomography (PET)[1] as a powerful non-invasive technique for investigating pathophysiology in living man.

PET is essentially an external technique for the rapid serial reconstruction of the spatial distribution of any positron-emitting radioisotope that has been administered in vivo. Such a distribution is primarily governed by the chemical form in which the positron-emitting radioisotope is incorporated, and importantly for clinical research, is often perturbed by physical, biological or clinical factors. Judicious choice of the chemical form enables specific biological. information to be obtained. For example, the labelling of glucose (1) with a positron-emitting radioisotope could be expected to provide a radiopharmaceutical for the study of glucose utilisation in both health and disease.

(1)

Glucose can be labelled with positron-emitting radioisotopes in several ways. Isotopic labelling, by substituting one or more stable carbons with ^{11}C, provides a true tracer of endogenous glucose. However, because glucose metabolism is rapid and complex, the position of label becomes important with respect to the intended application. Site specific labelling can be achieved chemically, if only at the C-1 position,[2] whereas general, though not necessarily uniform, labelling can be achieved photosynthetically.[3,4]

Rapidly metabolised tracers, like labelled glucose, are difficult to interpret in PET studies, because the signal (radioactivity) cannot easily be assigned to a chemical form. This difficulty is sometimes alleviated by labelling an analogue that has more simple and better understood biological behaviour. For example, replacement of the 2-OH group by H in glucose modifies its biological behaviour, preventing metabolism beyond initial 6-phosphorylation and causing entrapment within glycolytic cells. Non-isotopic labelling of 2-deoxy-glucose, with ^{18}F in the 2-deoxy position, provides a valuable radiopharmaceutical[5,6] for measuring tissue glucose uptake with PET. Here ^{18}F behaves only as a radioactive isostere of H and alters little the biological behaviour of the parent molecule; sometimes, however, replacement of H by F causes a profound change in biological behaviour,[7-9] especially when F is able to exert a strong electronic effect. It is also possible to label 2-deoxy-glucose isotopically with [11]C.[10] However, given the availability of 2-[^{18}F]fluoro-2-deoxy-glucose (^{18}F]FDG), the shorter half-life of [11]C (20.4 min) compared to that of [18]F (110 min) would needlessly restrict useful data acquisition in PET.

By contrast to ^{11}C]glucose, 2-deoxy-^{11}C]glucose or ^{18}F]FDG, 3-O-^{11}C]methyl-glucose[11] is regarded as non-metabolisable, but nonetheless provides a valuable radiopharmaceutical for measuring glucose transport.[12] Thus tracer variation can be invaluable to the dissection of physiological and biochemical processes with PET.

The chemical form for incorporation of a positron-emitting radioisotope need not, of course, be endogenous; it can, for example, be a proprietory drug, a receptor ligand, an enzyme inhibitor, an antibody fragment, or an analogue of any of these. PET might then be able to measure the distribution of the biological target (lesion, receptor, enzyme, antigen etc.) to which the particular exogenous species binds, thereby opening up the prospect of relating abnormal target distributions to certain diseases. Thus, for example, the antipsychotic drug and receptor ligand, spiperone (2), has been labelled by ^{11}C]methylation[13] at the amido-N atom and applied to study the role of brain dopamine receptor (type 2) concentration in schizophrenia.[14]

(2)

Such radioligands must be produced in high[15,16] and known specific activity (high radioactivity per unit mass of ligand). Otherwise stable ligand will occupy the always small number of target receptors at the expense of radioligand, thereby reducing the useful signal (receptor-bound radioactivity) for detection by PET. Radioligands must often possess other properties to work successfully, including high receptor affinity, high receptor selectivity, suitable lipophilicity, low toxicity, slow metabolism and, if targeted within brain, ability to cross the blood-brain-barrier.

Clearly, the diversity of clinical studies that can be contemplated with PET depends on the availability of suitable radiopharmaceuticals. The preceding examples serve to show that the main constraints on the design of positron-emitting radiopharmaceuticals are biochemical, physico-chemical, pharma-cological and toxicological. Versatile radiochemistry must exist if there is to be a good chance of complying with any particular set of constraints. The aim of this chapter is to introduce the reader to the current scope for developing radiopharmaceuticals for PET by considering in turn the physical properties, production and chemistry of the principal positron-emitting radioisotopes of the non-metals. Emphasis is placed on essentials and the major developments that have taken place since an earlier comprehensive review of this area.[17]

2. PHYSICAL PROPERTIES OF POSITRON-EMITTING RADIOISOTOPES

A positron (β^+) has the same rest mass as an electron but a positive charge. It is regarded as the anti-matter counterpart to an electron. An unstable neutron-deficient nucleus of low atomic number usually decays to a more stable nucleus by one of two main mechanisms, namely positron emission or electron capture.

Emission of a positron from a neutron-deficient nucleus effectively creates a neutron from a proton e.g.

$$^{11}_{6}C \longrightarrow ^{11}_{5}B + \beta^{+} \qquad (1)$$

The ejected positron has kinetic energy that is on average about one third of a maximal value, characteristic of the parent radioisotope (Table 1). This kinetic energy carries the positron through matter some distance that depends on local density. Maximal distances are of the order of mm in water (Table 1) or tissue. When the positron is almost at rest it annihilates with an electron and sends out two penetrative γ-rays of equal and specific energy (511 keV - equivalent to the rest mass of an electron) in almost exactly opposite directions (to conserve near zero momentum) (Figure 1).

Table 1.   Longer-lived Positron-emitting Radioisotopes of Non-metals

Isotope	β^{+} Emission[a] (%)	Maximal β^{+} energy (MeV)	Maximal β^{+} range[b] (mm in H_2O)	Half-life (min)	Decay product
^{11}C	99.8	0.96	4.1	20.4	^{11}B
^{13}N	100	1.19	5.4	9.96	^{13}C
^{15}O	99.9	1.723	8.3	2.03	^{15}N
^{18}F	96.9	0.635	2.4	109.8	^{18}O
^{30}P	100	3.245	16.8	2.5	^{30}Si
^{34m}Cl	53	2.47	12.5	32.0	^{34}S
^{73}Se	65	1.68	8.2	430.8	^{73}As[c]
^{75}Br	75.5	1.74	8.4	98.0	^{75}Se[c]
^{76}Br	57	3.98	20.8	966	^{76}Se
^{122}I	77	3.12	16.0	3.6	^{122}Te

[a]Remainder by electron capture, except ^{34m}Cl (internal transition).
[b]Interpolated from data in reference 18. Other nuclear data are from reference 19.
[c]Unstable.

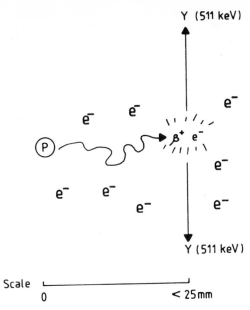

Figure 1.   Decay of a positron-emitting radionuclide (P)

It is this special feature of positron-electron annihilation, the simultaneous emission of two 511 keV γ-rays in opposite directions, that endows PET with quantitative accuracy and spatial resolution.[20]

For illustration, consider a disperse positron-emitting source (Figure 2).   A pair of small γ-ray detectors, placed on opposite

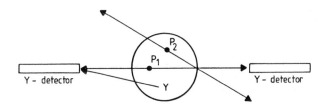

Figure 2.   Detection capability of a pair of γ-detectors set to record only coincident γ-rays from a disperse positron-emitting source. P_1-detected positron, P_2-undectected positron, γ-undetected gamma ray (from, for example, electron-capture)

sides of the source and set electronically to record only
coincident pairs of γ-rays, can only detect positron annihilation
along the axis between the detectors. This amounts to <u>electronic</u>
<u>collimation</u> and dispenses with any need for physical <u>collimation</u>
(e.g. thick lead shielding). So it becomes possible to surround a
positron-emitting source with adjacent rings of small electronic-
ally-collimated detectors, each able to respond in tandem with any
other to coincident γ-rays. Computerised analysis of the acquired
emission data can then give the true spatial distribution of
positron-electron annihilations encircled by the detectors (after
applying appropriate corrections for γ-ray attenuation).[20] This
distribution closely approximates to the distribution of positron-
emitting radioisotope, with the closeness of approximation
ultimately limited by positron range. In effect each ring of
detectors provides a scan that is equivalent to a quantitative <u>in</u>
<u>vivo</u> autoradiograph. Modern PET cameras can rapidly (e.g. every
20 s) construct scans from several adjacent planes and so provide
kinetic as well as structural or functional information.

In electron capture an unstable neutron-deficient nucleus
gains a neutron by capturing an orbital electron. The energy
change involved in this transition is expressed by the emission of
one or more γ-rays of specific energy, e.g.

$$^{11}_{6}C \longrightarrow {}^{11}_{5}B + \gamma \tag{2}$$

The relative proportions of decay by positron emission, electron
capture or other mode are fixed for a particular positron-emitting
radioisotope (Table 1). Electron capture is worthless to PET,
since no pair of γ-rays is generated for coincident detection.
Indeed, electron capture is a disadvantage since it contributes to
radiation dose but not to acquired data.

Amongst the non-metals, the half-lives of even the longest-
lived positron-emitting radioisotopes are very short, ranging from
2.03 min for ^{15}O to 966 min for ^{76}Br (Table 1). Most of these
radioisotopes give stable decay products (Table 1). For any such
radioisotope the radioactivity per Avogadro number of atoms is
inversely related to half-life ($t\frac{1}{2}$) and is given by (6.95 x
10^{21})/$t\frac{1}{2}$ Bq, where $t\frac{1}{2}$ is in minutes. Typically 370 MBq of radioactivity suffice for a single PET study in man; this
radioactivity represents very small quantities of mono-labelled
compound (e.g. 0.53, 1.08 and 5.83 pmol for ^{13}N-, ^{11}C- and ^{18}F-
labelled compound, respectively). Thus the short half-lives of
positron-emitting radioisotopes imply a possibility for very high
detection sensitivity in PET.

In reality a labelled compound is invariably found to be
diluted with the non-labelled compound or <u>carrier</u>; a truly
<u>carrier-free</u> (CF) state is seldom attained. A labelled compound

is described as <u>no-carrier-added</u> (NCA) if the CF state is unproven, but all <u>reasonable</u> precautions have been taken against dilution with carrier. If the labelled compound has been diluted deliberately with carrier then the labelled compound is described as <u>carrier-added</u> (CA).[21]

For a CF radioactive compound the specific activity is constant. However even a small dilution with carrier (e.g. 1%) causes the specific activity of a radioactive compound to decrease significantly over a few half-lives. At greater than 10-fold dilution with carrier the decrease in specific activity is virtually mono-exponential, with 't½' almost equal to the physical half-life of the radioisotope (Figure 3). Carrier dilutions are invariably greater than 10 for compounds labelled with positron-emitting radioisotopes. Thus, because these radioisotopes are very short-lived, values of activity and specific activity must relate to a particular time. In this chapter all such values relate to the end of radioisotope production (commonly known as end of bombardment or EOB). Correspondingly, all radiochemical yields are corrected to EOB. Where carrier dilution is known to be large (e.g. > 100-fold) it is often more convenient to talk about the carrier content (a constant) rather than the specific activity (a variable) of a given short-lived radioactive sample.

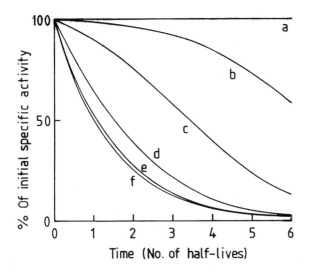

Figure 3. Change in specific activity for a radioactive compound with time, when initially a) CF, b) diluted 1% with carrier, c) 10% diluted with carrier, d) diluted 100% with carrier, e) diluted 10-fold and f) diluted infinitely with carrier

The physical properties of positron-emitting radioisotopes have important implications for radiochemists. Foremost is that any synthesis, including product purification and formulation for in vivo administration, must be completed rapidly, often within 3-4 half-lives of radioisotope production. Also thick (50-100 mm) lead-shielding is required to protect against γ-rays from high initial activities. This in turn demands that organic radiochemistry be remotely-controlled. In some ways the short-lived positron-emitting radioisotopes are easier to work with than the more familiar longed-lived β-emitters. Thus they are directly detectable with all radiation-sensitive devices. Moreover radioactive spillage or contamination decays to insignificance over short periods, enabling apparatus to be cleaned and re-used quickly.

Clearly, from the viewpoint of application in PET, the ideal physical properties required of a positron-emitting radioisotope are i) pure positron emission, ii) low positron energy, iii) a suitable half-life for data acquisition and iv) a stable decay product. From the viewpoint of radiopharmaceutical development the radioisotope should be adequately long-lived and easy to produce in i) high activity, ii) high specific activity, iii) high radionuclidic purity and iv) a major simple chemical form for chemical transformation. The radioisotopes that most nearly meet these two sets of requirements are ^{11}C, ^{13}N and ^{18}F (Table 1) and it is with these radioisotopes that the remaining parts of this chapter are concerned. Of the others only ^{75}Br and ^{76}Br (Ref. 22,23) have received more than scant attention for preparing radiopharmaceuticals.[24-26] Only a few of these have been found useful in PET, most notably [^{76}Br]bromospiperone for investigating dopamine receptors[27] and 15-(p-[^{75}Br]bromophenyl)pentanoic acid for investigating myocardial metabolism.[28]

3.   PRODUCTION OF ^{11}C, ^{13}N AND ^{18}F FOR ORGANIC SYNTHESIS

Though many devices and methods exist for the production of radioisotopes in general, cyclotrons (and other charged-particle accelerators) have become, for good physical reasons, the most popular sources of positron-emitting radioisotopes. Several models of cyclotron are now available commercially. Wolf and Jones have usefully classified these cyclotrons as Level I, II, III or IV according to the energy of protons and multiplicity of particles on offer.[29] Most commercial cyclotrons are capable of worthwhile ^{11}C, ^{13}N and ^{18}F production because the most useful nuclear reactions have low threshold energies and use either protons or deuterons as incident particle (Table 2). Many commercial cyclotrons at Level I (< 10 MeV p or d) or Level II (< 20 MeV p, plus perhaps other particles) are specifically intended to support PET programmes.

Table 2. Common Methods for the Production of ^{11}C, ^{13}N, and ^{18}F

Method	Threshold energy (MeV)	Main product(s)	Typical yield (GBq)	Typical specific activity (GBq/µmol)	Dilution with carrier
$^{14}N(p,\alpha)^{11}C$ on N_2	3.13	$^{11}CO_2$	93[a]	150	2 x 10³
$^{14}N(p,\alpha)^{11}C$ on N_2(5% H_2)	3.13	$^{11}CH_4$	67[b]	130	3 x 10³
$^{12}C(d,n)^{13}N$ on CH_4	0.3	$^{13}NH_3$	2.2[c]	0.02	4 x 10⁷
$^{16}O(p,\alpha)^{13}N$ on H_2O	5.5	$^{13}NO_3^-$ $^{13}NO_2^-$	18.5[d]	> 370	< 2 x 10³
$^{20}Ne(d,\alpha)^{18}F$ on Ne(0.1% F_2)	0	^{18}F-F	13.6[e]	~ 0.2	~ 3 x 10⁵
$^{16}O(^3He,p)^{18}F$ on H_2O	0	$^{18}F^-(H_2O)_n$	14.8[f]	138	5 x 10²
$^{18}O(p,n)^{18}F$ on $H_2^{18}O$	2.5	$^{18}F^-(H_2O)_n$	44[g]	370	2 x 10²

[a] 19 MeV, 30 µA, 30 min.   [b] 18 MeV, 30 µA, 40 min.[33]   [c] 8 MeV, 15 µA, 10 min.[38]   [d] 19 MeV, 15 µA, 20 min.[40]   [e] 14 MeV, 15 µA, 120 min.[41]   [f] 36 MeV, 40 µA, 60 min.[43]   [g] 15 MeV, 20 µA, 60 min.[44]

For any method of radioisotope production the range of projectile energy that gives the best yield of the desired radioisotope in an acceptable radionuclidic purity is of paramount importance. Extensive data on the cross sections of nuclear reactions, including excitation functions, have been accumulated for this purpose and the status of this data has been reviewed quite recently.[30]   In theory, at least, radioisotope yield is proportional to beam current.   Usable beam current is seldom limited by the available cyclotron but by targetry considerations;   a target irradiated with 20 MeV protons at only 10 μA must cope with a considerable influx of energy (200 W).   The Achilles heel is often the target window, usually a thin metal foil, that may be vulnerable to rupture from thermal damage, pressure stress, corrosion or radiation damage (Figure 4).   Good target design is thus fundamental to efficient and reliable radioisotope production.   The proceedings of a recent workshop[31] reveal the 'state-of-the art' in this important area.

The production of positron-emitting radioisotopes has been reviewed extensively.[22,32]   Brief descriptions of only the most general and useful methods of producing ^{11}C, ^{13}N and ^{18}F follow here (Table 2).

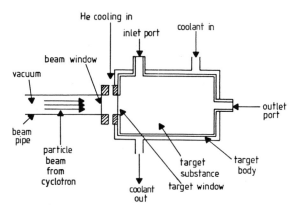

Figure 4.   A schematic representation of a cyclotron gas target

Production of ^{11}C

The most widely used method for ^{11}C production is the $^{14}N(p,\alpha)$ ^{11}C reaction[33-35] on nitrogen gas (Table 2).   High activities can be produced.   The chemical form of the obtained ^{11}C depends on target composition.   High purity nitrogen (with trace oxygen) provides mainly [^{11}C]carbon dioxide, with [^{11}C]carbon monoxide as radiochemical impurity.   With hydrogen (5%) present, and with adequate irradiation (0.1 eV/molecule), mainly [^{11}C]-methane is obtained.[36]   Nevertheless   most   centres   prefer   to   prepare [^{11}C]methane by Ni-catalysed reduction of [^{11}C]carbon dioxide with hydrogen at 400°C.   This can be performed 'on-line' with high efficiency.

Because the generated ^{11}C is diluted within the target with a large volume of nitrogen, pre-concentration of the ^{11}C is often necessary before use.   [^{11}C]Carbon dioxide may be pre-concentrated by passage into molecular sieve (4 Å) at room temperature or into a stainless steel tube immersed in liquid argon.[37]   [^{11}C]Methane may be pre-concentrated by passage into Porapak Q at -180°C.   In each case radioactivity is recovered by passing nitrogen (or helium) through the heated trap (20-100°C for stainless steel loop, > 220°C for molecular sieve and RT for Porapak Q).   These pre-concentration methods separate off [^{13}N]nitrogen arising from the ^{14}N(p,pn)^{13}N reaction.   Both production methods achieve high specific activities if care is taken to exclude extraneous carbon dioxide from the target system (Table 2).[37]

Production of ^{13}N

There are two main procedures that produce ^{13}N for radiochemistry, the ^{12}C(d,n)^{13}N reaction on methane[38] and the ^{16}O(p,α)^{13}N reaction on water[39] (Table 2). Deuteron irradiation of methane produces mainly [^{13}N]ammonia but yields are low. Moreover radiolysis of trace nitrogen, which is difficult to eradicate from methane, leads to stable ammonia and very low specific activity (Table 2).   Radiolysis of the methane itself gives a copious quantity of oil which is troublesome in radionuclide recovery and in subsequent chemistry.

Proton irradiation of deaerated water is the preferred method for NCA ^{13}N production.   Optimal production yields are much higher than for the ^{12}C(d,n)^{13}N process (Table 2). ^{13}N is mostly obtained as a mixture of [^{13}N]nitrate and [^{13}N]nitrite at good specific activity[40] (Table 2). ^{15}O and ^{18}F (as fluoride) appear at low levels, but these do not generally interfere with subsequent chemistry.

Production of ^{18}F

Several methods for the production of ^{18}F are in common use.

CA ^{18}F is prepared by the deuteron irradiation of neon with added fluorine (ca. 0.1% v/v) at a pressure up to 25 bar.   Useful activities can be produced from high beam currents (50 µA) of low energy deuterons (10 MeV) or, more usually, lower beam currents of higher energy deuterons (Table 2).   Detailed studies[41,42] have shown that target construction and gas composition are important with regard to obtaining a high recovery of radioactivity as molecular [^{18}F]fluorine. Monel or nickel targets, passivated with fluorine (1%) during irradiation, give good recovery, provided that target gases are kept free of contaminants, especially nitrogen, carbon dioxide and carbon tetrafluoride.

NCA ^{18}F can be produced by irradiating water with high energy (> 24 MeV) α-particles, or, for higher yield, lower energy ^{3}He^{2+} (Table 2).[45,46] Recently, even greater yields of NCA ^{18}F have been produced by irradiating ^{18}O-enriched water with moderate energy protons.[47] Targets that use less than 1 ml of ^{18}O-enriched water have been developed.[47,48] The major chemical form from these processes is [^{18}F]fluoride, i.e. ^{18}F$^-$(H_2O)$_n$. Though ^{18}O-enriched water is expensive, it can be re-used after distillation or after recovery of the [^{18}F]fluoride by ion exchange. Target materials are important with respect to radionuclidic purity and to reactivity of the [^{18}F]fluoride. Materials containing Cr, Fe or Co are undesirable, as they possibly produce unreactive metal [^{18}F]fluorides, but Ni, Ti and Ag are acceptable. Very high specific activities can be achieved from water targets, provided that fluoride-free water is used (Table 2).

4. ORGANIC CHEMISTRY WITH ^{11}C, ^{13}N and ^{18}F

Chemistry with ^{11}C

In principle ^{11}C is a candidate for the isotopic or non-isotopic labelling of any organic compound (endogenous or exo-genous) and is thus of fundamental importance to bio-medical studies with PET. ^{11}C can be produced in high initial activity (Table 2) and this compensates for its short half-life in organic synthesis. All methods of ^{11}C production, result in some dilution with carrier. For the main methods, this dilution, though large (Table 2), still permits the possibility of preparing radioligands in adequate specific activity (generally > 37 GBq/μmol). Some-times, in practice, the use of a high initial activity of ^{11}C is solely justified by the need to obtain high specific activity in the desired product.

Chemistry with ^{11}C stems from the chemistry of the primary irradiation products, [^{11}C]carbon dioxide and [^{11}C]methane. Only [^{11}C]carbon dioxide is directly useful for the syntheses of radio-pharmaceuticals (Figure 5). Thus [1-^{11}C]carboxylates can be produced in high radiochemical yields *via* [^{11}C]carbonations of Grignard reagents.[49] Important examples are the preparations of [1-^{11}C]acetate[50] and [1-^{11}C]palmitate[51,52] for studies of myocardial fatty acid metabolism. Many α-lithioisonitriles can be carbonated to give D,L-[carboxy-^{11}C]amino acids upon hydroly-sis.[53,54] The fast resolution of such racemic amino acids can be approached enzymically[55], by chiral HPLC[56] or by selective binding

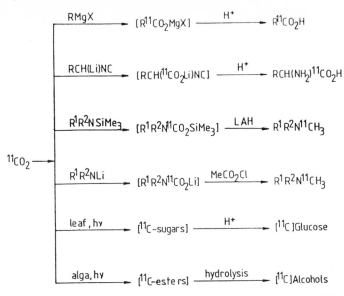

Figure 5.   Radiopharmaceutical preparation from [^{11}C]carbon dioxide

to human serum albumin.[57] Labelled L-amino acids are of potential value to oncology with PET. Recently the scope of [^{11}C]carbonation has been extended to the N-trimethylsilyl and N-lithio-derivatives of a secondary amine in order to prepare the [N-methyl-^{11}C]tertiary amine, [^{11}C]imipramine (3),[58] a radioligand for serotonergic receptors, and also to α-lithiopyrrolidyl-N-tert-butyl-formamidine for the synthesis of D,L-[1-^{11}C]proline.[59] [^{11}C]Carbon dioxide has also been used directly in photosynthesis,

(3)

for example, by plant leaves[60,61] to produce [11C]glucose, and by blue-green alga[62] to produce [11C]mannitol or [11C]glycerol. Nonetheless, the vast majority of syntheses with 11C require a primary irradiation product to be converted into a suitable labelling agent. A wide variety of such labelling agents can now be prepared (Figure 6), often efficiently (Table 3) and with specific activity adequate for the preparation of radioligands.

[11C]Iodomethane was early developed as a labelling agent[63,64] and remains the most widely used. With this agent [11C]methylations have been achieved at C, N, O, P, S and Se atoms in a wide range of structures, with those at N atoms in amines and amides most numerous.[17]

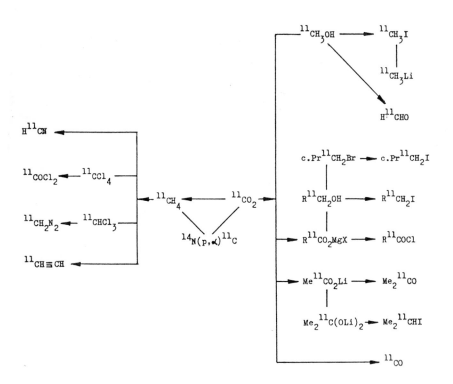

Figure 6. Routes to 11C-labelling agents from cyclotron-produced [11C]carbon dioxide and [11C]methane

Table 3. Typical Production Parameters for ^{11}C-Labelling Agents

Labelling agent	Preparation time (min)	Radiochemical yield[a] (%)	Production efficiency[b] (%)
H^{11}CN	5	95	80
^{11}CO	5	90	76
^{11}CH$_3$I	10	80	57
^{11}CH$_3$Li	15	90	54
Me$_2$^{11}CO	5	57	48
^{11}CH≡CH[c]	9	60	44
H^{11}CHO	8	55	42
R^{11}COCl[d]	10	15-60	10-40
R^{11}CH$_2$I[e]	10-14	13-52	8-37
^{11}COCl$_2$	10	42	30
Me$_2$^{11}CHI	12	36	24
^{11}CH$_2$N$_2$	10	31	22
c.Pr^{11}CH$_2$I	20	10	5

[a] Decay-corrected to EOB from irradiation product.

[b] Non decay-corrected radiochemical yield.

[c] CA preparation, giving a specific activity of 5 GBq/μmol; other preparations are NCA and give specific activities in the range 40-180 GBq/μmol.

[d] R = Me, c.Pr, 2-furoyl.

[e] R = Et, Pr, i-Pr, Ph.

[^{11}C]Iodomethane is usually prepared from cyclotron-produced [^{11}C]carbon dioxide by reduction with lithium aluminium hydride (LAH), followed by treatment with hydroiodic acid (HI) i.e.

$$^{11}CO_2 \xrightarrow[\text{ii) } H_2O]{\text{i) LAH}} {}^{11}CH_3OH \xrightarrow{\text{HI}} {}^{11}CH_3I \qquad (3)$$

In order to minimise one source of carrier (i.e. CO_2) it is essential to prepare and handle the LAH under an inert atmosphere. For the same reason it is essential to use the smallest volume and quantity of reagent that will still give efficient reduction.[37,65] These principles apply to all reagents that are used to prepare [^{11}C]labelling agents from [^{11}C]carbon dioxide. In practice, optimal reaction volumes are often as small as 0.1-1.0 ml and reagent concentrations ~ 0.1M. Carrier may be inadvertently introduced at any stage of a synthesis. For example, it has even been suggested that tetrahydrofuran, the usual solvent for the LAH reduction of [^{11}C]carbon dioxide, can degrade to give carrier methanol.[65] For high specific activity it is imperative that all reagents are as free as possible from reaction products. In particular the compound to be labelled is often a convenient source of the substrate to be reacted with the labelling agent; if high specific activity is required the substrate must be purified to remove all trace of target compound.

Obviously, in any reaction between a stable compound and a labelling agent, the labelling agent (active plus stable) must not be in excess. Otherwise the potential radiochemical yield is reduced. For efficient [^{11}C]methylation of a primary or secondary amine, it is also necessary to have a large excess of amine over [^{11}C]iodomethane. This not only promotes rapid incorporation of radioactivity into product but also minimises loss of radioactivity by sequential reaction to unwanted dimethyl product e.g.

$$RNH_2 \xrightarrow{^{11}CH_3I/MeI} RNHMe/RNH^{11}CH_3 \xrightarrow{^{11}CH_3I/MeI} RN(Me)^{11}CH_3 \qquad (4)$$

The amount of amine that may be used is often limited to ~ 20 µmol by the capacity of the subsequent purification procedure (usually semi-preparative HPLC) and possibly by the need to conserve precious substrate. This generally gives a greater than 50-fold excess of amine over iodomethane (stable plus active). Though the generated N-methylamine is only ever present at low concentration, compared to the nor-compound, its sometimes greater reactivity may still give an appreciable radiochemical yield of undesired dimethyl product.[66] Formation of the desired product may also be hampered by competing reactions of the labelling agent with solvent (e.g. $^{11}CH_3I$ with DMSO), traces of degraded solvent or contaminant. A small solvent volume (e.g. 0.1-1.0 ml) reduces these problems and promotes fast product formation.

With a large excess of substrate over labelling agent, reaction predominates at the most reactive centre. For reaction at a less reactive centre, other reactive groups need protection. Rapid and efficient deprotection must then be possible. A simple example is the preparation of 3-O-[^{11}C]methyl-glucose via the [^{11}C]methylation of 1,2,5,6-diacetone glucose in which deprotection is achieved by brief acid hydrolysis.[11] Synthon chemistry has been exploited to achieve reaction at non-reactive sites[67], sometimes enantioselectively.[68] Again the labelling reaction is followed, by other fast, usually hydrolytic, reactions.

Frequently, in the development of radiopharmaceuticals, it is desirable to optimise the non-decay-corrected radiochemical yield of a reaction with respect to reaction time, solvent, substrate concentration and temperature. Multivariate strategies[69] such as Simplex,[70] have been proposed[71] as efficient means of optimisation. Of course, formation of radioactive product competes with loss by physical decay; therefore optimal reaction times are often much less than one half-life.

The overall non-decay-corrected radiochemical yield of a preparation represents its overall 'efficiency' (the percentage of initial radioactivity incorporated into final product). Clearly the maximal theoretical efficiency halves for every half-life taken for the preparation. For the sake of overall efficiency all reactions and product transfers should be individually efficient and their total number kept to a minimum. For example, one reaction plus one transfer, each at 50% efficiency, give an overall efficiency of 25%, while two reactions and two transfers, each at 70% efficiency, give an overall efficiency of only 24%. These considerations are not trivial in practice; even the simplest preparation of a labelled compound by [^{11}C]methylation involves ten stages (reactions or transfers) after EOB. If the overall efficiency of a preparation is more than adequate (> 2% for ^{11}C), some may be usefully sacrificed for improved specific activity by simply reducing reaction times.

These comments on the control of carrier dilution, reaction stoichiometry, regioselectivity and optimisation of yield and specific activity apply in an obvious sense to the chemistry of other [^{11}C]labelling agents. (Comments on yield optimisation also apply to chemistry with ^{13}N and ^{18}F). Hence only specific features of the preparation and use of other [^{11}C]labelling agents will now be discussed.

[^{11}C]Formaldehyde, in reductive [^{11}C]methylation, is an alternative agent to [^{11}C]iodomethane for N-[^{11}C]methylation, but is seldom used because its production via high temperature (ca 350°C) Ag-catalysed oxidation of [^{11}C]methanol[63,72,73] depends on the activation of the catalyst and is poorly reproducible. It is, however, sometimes useful for selective [^{11}C]methylations on sensitive substrates[74,75] and may succeed where [^{11}C]iodomethane fails to give adequate reaction.[76]

[^{11}C]Methylations have been achieved at electron-deficient centres using [^{11}C]methyl lithium, which is itself prepared from [^{11}C]iodomethane with butyl lithium.[77]

Higher [^{11}C]iodoalkanes ([1-^{11}C]EtI,[78-80] [1-^{11}C]PrI,[80] [1-^{11}C]BuI,[80] [1-^{11}C]iso-BuI,[80] [2-^{11}C]iso-PrI[81] and [1-^{11}C]iodo-methylcyclopropane[82]) and also [1-^{11}C]benzyl iodide[83] have been developed and applied as alkylating agents. The primary acyclic [1-^{11}C]iodoalkanes and [1-^{11}C]benzyl iodide are prepared via [^{11}C]carbonation of a Grignard reagent, reduction with LAH and treatment with HI i.e.

$$RMgBr \xrightarrow{\text{\ \ }^{11}CO_2\text{\ \ }} [R^{11}COOMgBr] \xrightarrow{LAH} R^{11}CH_2OH \xrightarrow{HI} R^{11}CH_2I \quad (5)$$

[1-^{11}C]Cyclopropane carbinol can be prepared along a similar route but treatment with HI causes rearrangement through a delocalised carbocation to give a mixture of isomeric [^{11}C]iodo-alkanes i.e.

$$(6)$$

This rearrangement is largely circumvented by converting the [^{11}C]alcohol into the [^{11}C]bromide with p-toluenesulphonyl bromide. The [^{11}C]bromide gives [1-^{11}C]iodomethylcyclopropane cleanly on brief treatment with sodium iodide in acetone.[82] The overall synthesis is however poorly efficient (Table 3). [^{11}C]Bromoalkanes are themselves too unreactive to be useful alkylating agents in ^{11}C-chemistry.

[1-^{11}C]Iodomethylcyclopropane has potential for introducing the N-[^{11}C]cyclopropylmethyl group into opiate receptor ligands, such as diprenorphine (4). Originally this was achieved in two steps : acylation of the appropriate amine with [1-^{11}C]cyclo-propanecarbonyl chloride, followed by LAH reduction:[84]

$$(7)$$

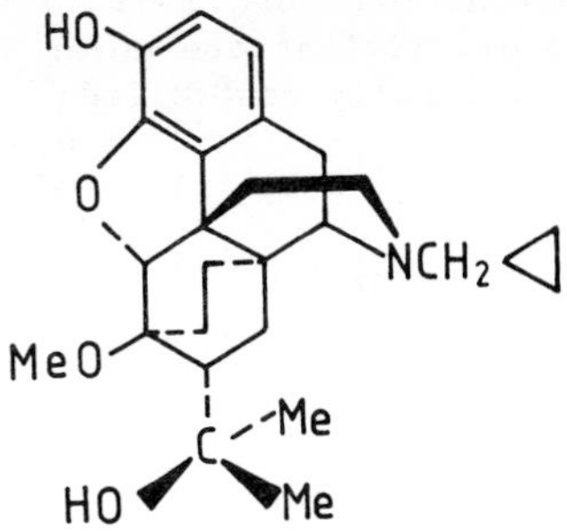

(4)

The [1-^{11}C]cyclopropanecarbonyl chloride is easily prepared by the [^{11}C]carbonation of cyclopropylmagnesium bromide, followed by treatment with phthaloyl dichloride (PDC):

$$RBr \xrightarrow{\;^{11}CO_2\;} [R^{11}COOMgBr] \xrightarrow{\;PDC\;} R^{11}COCl \qquad (8)$$

Other volatile [1-^{11}C]acid chlorides have been prepared similarly and used to label compounds of bio-medical interest, including the α_1-receptor ligand, prazosin (5), with [1-^{11}C]furoyl chloride[85] and the neurohormone, melatonin (6), with [1-^{11}C]acetyl chloride.[86]

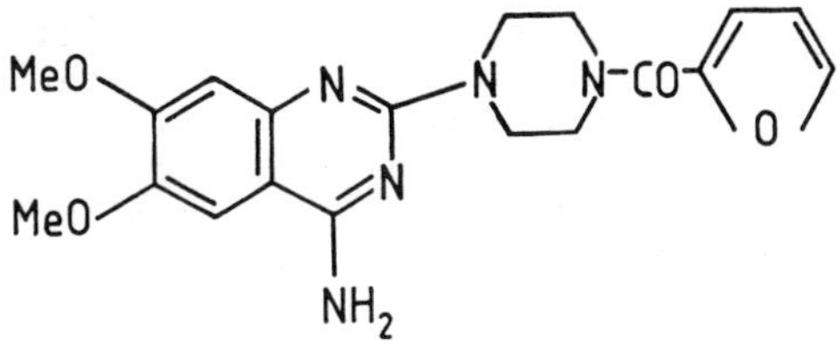

(5)

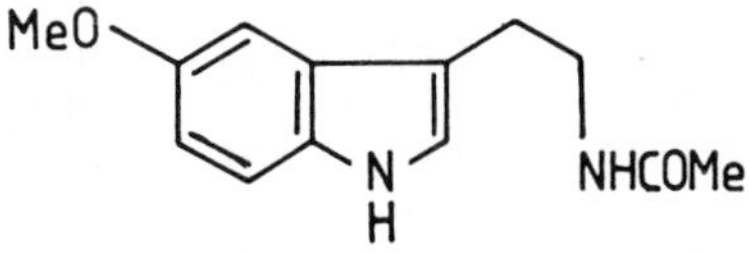

(6)

The [^{11}C]carbonation of methyllithium is a route to [2-^{11}C]acetone, provided that the molar ratio of methyllithium to carbon dioxide is carefully controlled:[87]

$$^{11}CO_2 \xrightarrow{\text{2 MeLi}} [Me_2{}^{11}C(OLi)_2] \xrightarrow{\text{H}_2\text{O}} Me_2{}^{11}CO \qquad (9)$$

Excess methyllithium gives [2-^{11}C]tert-butanol as major product:

$$[Me_2{}^{11}C(OLi)]_2 \xrightarrow[\text{H}_2\text{O}]{\text{MeLi}} Me_3{}^{11}COH \qquad (10)$$

[2-^{11}C]Acetone is particularly useful for labelling b-receptor ligands in their N-isopropyl groups by reductive alkylation.[88-90]

Reduction of Me$_2$^{11}C(OLi)$_2$ with LAH followed by treatment with HI provides [2-^{11}C]iodo-isopropane.[81] This has been used to prepare D,L-[3-^{11}C]valine *via* alkylation of a glycine synthon.[81]

[^{11}C]Carbon monoxide is easily prepared by high temperature (390°C) Zn-catalysed reduction of [^{11}C]carbon dioxide,[91] but is little explored as a labelling agent. Some work however shows the potential of [^{11}C]carbon monoxide for labelling amides[92] and aldehydes.[93] [^{11}C]Carbon monoxide has also served as an intermediate in early routes[94,95] to another labelling agent, [^{11}C]phosgene. These routes caused some dilution with carrier, though radioligands[96,97] could still be prepared in acceptably high specific activity. A recently developed route for [^{11}C]phosgene from [^{11}C]methane, *via* chlorination to [^{11}C]carbon tetrachloride and Fe-catalysed oxidation, gives 3-fold better specific activity.[98]

The chlorination of [^{11}C]methane can be controlled to give [^{11}C]chloroform, which in turn may be converted into [^{11}C]diazomethane of high specific activity by reaction with hydrazine and potassium hydroxide.[99] [^{11}C]Diazomethane has been used successfully to prepare a [^{11}C]methyl ester,[99] but remains to be more fully explored as an alternative to [^{11}C]iodomethane.

Pyrolysis of [^{11}C]methane in an argon plasma produces [^{11}C]acetylene in good radiochemical yield, but of course with only modest specific activity.[100] CA [^{11}C]acetylene, produced by other methods has been used to label the steroids, moxestrol (7)[101] and 17-a-ethynylestradiol (8).[102]

The passage of [^{11}C]methane with ammonia over hot (1000°C) platinum gives hydrogen [^{11}C]cyanide in high specific activity and very high radiochemical yield.[36,103] Hydrogen [^{11}C]cyanide is used mainly to prepare natural[104] and cyclic[105,106] D,L-[carboxy-^{11}C]amino acids *via* the Bucherer-Strecker reaction (reaction of cyanide with aldehyde or ketone plus ammonium salt to give hydantoin, followed by alkaline hydrolysis). Bisulphite adducts

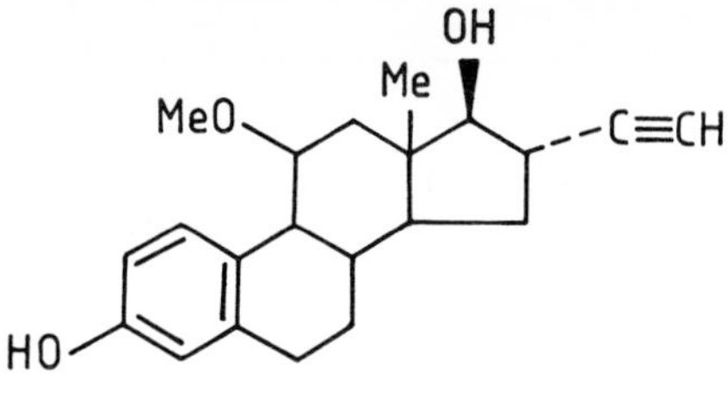

OH

Me

MeO ... $-C\equiv CH$

HO

(7)

OH

Me

$-C\equiv CH$

HO

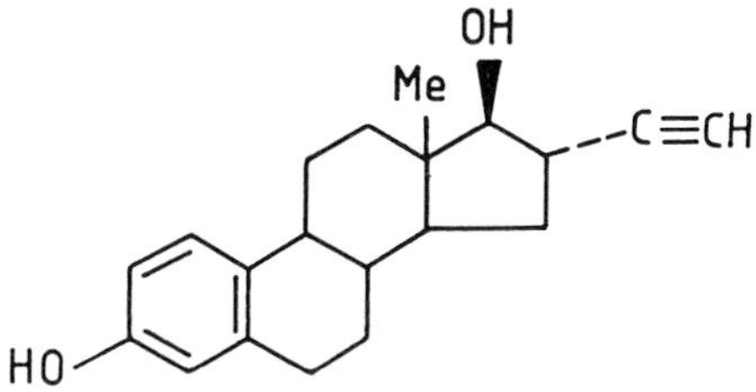

(8)

of the ketone or aldehyde may also be used.[103,107] Hydrogen [11C]cyanide is also important for the preparation of [11C]nitriles by nucleophilic substitutions. Such nitriles may then be reduced to [11C]amines.[108]

With the availability of such a broad range of labelling agents, most organic compounds are now amenable to isotopic labelling with [11]C. Moreover there is often considerable freedom to label a target molecule at a desired position rather than at a position dictated solely by chemical feasibility.

Chemistry with [13]N

Though ^{13}N has a rather short half-life for extensive chemistry, it has a special role as the only radioactive tracer of nitrogen. As such it has long been applied to tracer studies in microorganisms, plants and mammals. The chemistry of ^{13}N has been extensively reviewed[109] quite recently and so only a brief summary is given here.

[13N]Ammonia from the deuteron irradiation of methane has been used in synthesis, but nearly all work is now carried out with ^{13}N from the higher yield proton irradiation of water. The mixture of radioactive products from this process are usually converted into a single chemical form for subsequent chemistry (Figure 7).

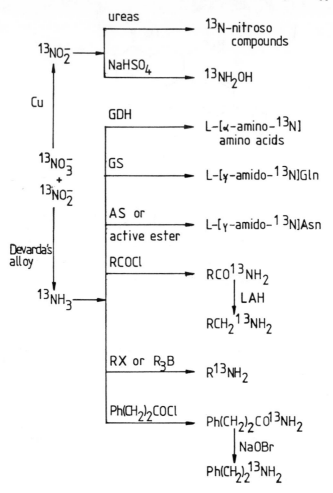

Figure 7. Chemistry with ^{13}N

Treatment of the irradiation products with carrier nitrate and Cu dust gives [^{13}N]nitrite.[110] This has been used to introduce the ^{13}N-nitroso group into nitrosoureas such as the carcinogen, nitrosocarbamoyl urea, and the carcinostat 1,3-bis-(2-chloroethyl)-1-nitrosourea.[111] [^{13}N]Nitrite has also been converted into hydroxyl[^{13}N]amine with sodium bisulphate.[112]

Most ^{13}N-chemistry uses [^{13}N]ammonia as labelling agent. This can be generated from proton-irradiated water by reduction with titanium(III) chloride,[113] zinc[114] or more usually Devarda's alloy in hot alkaline solution.[39,115] Devarda's alloy is known to cause some dilution of specific activity.[116] Such dilution is reduced considerably if the alloy is pre-heated at 170° for 24 hours.[116] With this method [13N]ammonia can be produced in ~ 90% radio-chemical yield within five minutes and with high specific activity. Furthermore, the evolved [13N]ammonia may be trapped in either aqueous or organic[117] solution for subsequent chemistry.

Several enzymes catalyse reactions of ammonia. Those that are commercially available were early exploited in rapid syntheses of L-[13N]amino acids. Of special importance has been L-glutamate dehydrogenase (GDH) which catalyses the amination of many α-keto acids:

$$RCO.CO_2H \; + \; NH_3 \; \longrightarrow \; RCH(NH_2)CO_2H \hspace{3cm} (11)$$

The wide substrate selectivity of this enzyme has been exploited to label several L-amino acids (e.g. Glu, Ala, Leu, Val, Met) with ^{13}N in useful radiochemical yields.[118-121] Radiochemical yields are noted to improve (e.g. to 95% for [13N]Glu) with an increase in the specific activity of the [13N]ammonia.[118,122]

L-Glutamine synthetase (GS) and L-asparagine synthetase (AS) have been used to label the amino groups of L-glutamine[118,122] and L-asparagine, respectively, as follows.

$$HO_2C(CH_2)_nCH(NH_2)CO_2H \; + \; ^{13}NH_3 \; \longrightarrow \; H_2{}^{13}NOC(CH_2)_nCH(NH_2)CO_2H) \hspace{0.5cm} (12)$$

Transaminases, such as L-glutamate-oxaloacetate transaminase, have also been used to convert [13N]glutamate into less directly accessible ^{13}N-labelled amino acids, such as L-[13N]aspartate[123] and L-[13N]tyrosine.[124]

The use of soluble enzymes in radiosyntheses may pose a hazard of pyrogenicity or even of antigenicity in the final product. Immobilised enzymes can avoid these problems and have the extra advantage of being re-usable.[119-125]

Non-enzymic methods of synthesis with [13N]ammonia have focussed on the preparation of [13N]amides and [13N]amines. [13N]Amides are generally prepared by rapid aminations of acid chlorides or anhydrides.[126-128] Syntheses are possible at the NCA level. Aminolysis of an activated ester may also be used. Thus [amido-13N]glutamine has been prepared by reaction of [13N]ammonia with the β-N-hydroxysuccimidyl ester of α-N-t-BOC-t-Bu-aspartate followed by acid hydrolysis.[129] The radiochemical yield (40%) exceeds that of the enzymic route.

[13N]Amines can be prepared directly via the [13N]amination of organoboranes[130] or halo compounds.[128] Addition of carrier limits unwanted sequential reactions of the type:

$$R^{13}NH_2 \xrightarrow{\text{RX}} R_2^{13}NH \xrightarrow{\text{RX}} R_3^{13}\overset{+}{N} \tag{13}$$

[^{13}N]Amines can also be prepared from [^{13}N]amides by LAH reduction[126] or by Hofmann rearrangement.[127] [^{13}N]Amination of phenylpropionyl chloride, followed by Hofmann rearrangement, was successful for the synthesis of [^{13}N]-β-phenethylamine at a NCA level of specific activity.[127]

Chemistry with ^{18}F

Fluorine is almost xenobiotic; certainly functional fluoro-organic compounds are unknown in man. Nonetheless ^{18}F is for many reasons a useful radioisotope for bio-medical studies. Physically, the longer half-life and shorter positron range of ^{18}F, compared to those of ^{11}C and ^{13}N (Table 1), provide more scope for data acquisition and better resolution in PET. Chemically, C-F bonds are stronger than C-H bonds and are generally stable in vivo. Exceptions are those in suicide inhibitors which are labile at the active sites of target enzymes.[131] Mono-fluoro compounds are often well-tolerated in vivo; indeed several of the first enzymes on the metabolic pathways of endogenous compounds accept a mono-fluoro derivative as a substrate (F is almost isosteric with H). Many such derivatives when labelled with ^{18}F, even at low specific activity, turn out to be useful radiopharmaceuticals (e.g. [^{18}F]FDG).

Sometimes the effects of replacing H by F are profound. F is highly electronegative and can affect neighbouring group pK_a or reactivity. F is also isoelectronic with OH and can participate in hydrogen bonding with H, sometimes perturbing molecular conformation. Replacement of H by F also increases lipophilicity and may alter bio-distribution. Since these effects are exploited in drug design an increasing number of drugs contain F in their structures. Some of these (e.g. spiperone (1)) are useful radio-ligands when labelled isotopically,[132] or even non-isotopic-ally,[133] with NCA ^{18}F.

Conveniently, most chemistry with ^{18}F is describable as either CA and electrophilic (that with [^{18}F]fluorine and its derivatives), (Figure 8) or NCA and nucleophilic (that with [^{18}F]fluoride).

CA and Electrophilic. Molecular fluorine is extremely reactive in oxidations, non-selective free radical reactions and electrophilic-type additions and substitutions. Regioselective fluorination of organic compounds is therefore difficult to achieve unless the fluorine is dilute and at low temperature. [^{18}F]Fluorine from the ^{20}Ne(d,α)^{18}F process contains carrier fluorine already diluted ~ 500-fold with Ne, and is sufficiently controllable for some applications.

1. Electrophilic addition

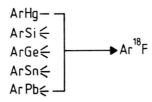

2. Electrophilic substitution

$$ArH \longrightarrow Ar^{18}F$$

3. Aryl-metal bond cleavage

ArHg—
ArSi≤
ArGe≤ $\longrightarrow Ar^{18}F$
ArSn≤
ArPb≤

Figure 8. Chemistry with ^{18}F-fluorine or ^{18}F-acetyl hypofluorite

Serious effort has been directed at developing other labelling agents from $[^{18}F]$fluorine. A recent advance has been the development of acetyl $[^{18}F]$hypofluorite as a milder 'electrophilic' labelling agent. This may be prepared in almost theoretical yield (50%) from the reaction of $[^{18}F]$fluorine with acetate salts in either the liquid[134,135] or gas phase[136,137]:

$$2^{18}F_2 + 2MeCOO^- \longrightarrow MeCOO^{18}F + MeCOOF + {}^{18}F^- + {}^{18}F^- \qquad (14)$$

Production of acetyl $[^{18}F]$hypofluorite in the gas phase enables a variety of solvents to be used in subsequent chemistry. Other labelling agents, prepared from $[^{18}F]$fluorine, suffer some major disadvantage and are now seldom used. Thus xenon $[^{18}F]$difluoride[138] is lengthy to prepare and perchloryl $[^{18}F]$fluoride is hazardous.[139] All of these reagents ($^{18}F_2$, MeCOO^{18}F, Xe$^{18}F_2$ and ClO$_3^{18}F$) can only at best introduce half of the initial radioactivity (initially $^{18}F_2$) into a mono-fluoro compound.

$[^{18}F]$Fluorine adds to double bonds, usually in a syn manner.[140] An important application has been the synthesis of $[^{18}F]$FDG by reaction of $[^{18}F]$fluorine with 3,4,6-tri-O-acetyl-D-glucose in freon, followed by hydrolysis. Addition to the least hindered face of the molecule is preferred. Thus $[^{18}F]$FDG and its 2-epimer are formed in the ratio 9:1.[141] The use of acetyl

[18F]hypofluorite under similar conditions increases this ratio to 19:1 and doubles the radiochemical yield of [18F]FDG to 20%.[142] Generally, however, the degree of stereoselectivity achieved with [18F]fluorine or acetyl [18F]hypofluorite is largely dictated by solvent[143,144] and substrate[145,146] and it is not possible to say that one reagent is more selective than the other.

Acetyl [18F]hypofluorite adds to a double bond to give an unstable intermediate that easily loses acetic acid in the presence of base to form an 18F-labelled vinyl fluoride i.e.

$$\underset{\text{H}}{\diagdown}\diagup \quad \xrightarrow{\text{AcO}^{18}\text{F}} \quad \overset{^{18}\text{F} \quad \text{OAC}}{\diagup\diagdown} \quad \xrightarrow{\text{base}} \quad \overset{^{18}\text{F}}{\diagdown}\diagup \qquad (15)$$

This type of process has proved useful for the 18F-labelling of 4-fluoroantipyrine (9)[147], a blood flow marker, and also of 5-[18F]-fluorouracil[148] and 5'-deoxy-5-fluoro-uridine[149] for PET studies in oncology.

(9)

Aromatic systems can sometimes be directly [18F]fluorinated to give low, but useful, yield of [18F]aryl fluorides. An interesting example is the direct [18F]fluorination of the anti-Parkinsonian drug, L-DOPA (L-3,4-dihydroxyphenylalanine), in anhydrous HF at -65°C, a reaction between a strong oxidising agent and an easily oxidisable substrate.[150] This gives a low yield of L-2-,5-, and 6-[18F]fluoro-DOPAs, in the ratio of 7:1:12. Here HF prevents excessive oxidation by protonating the hydroxy groups. Recycle HPLC can give a 3% radiochemical yield of pure L-6-[18F]-fluoro-DOPA, the required isomer for PET studies of brain dopamine turnover in Parkinson's disease and other movement disorders. Acetyl [18F]hypofluorite reacts with N-acetyl-3-0-methyl-4-0-acetyl-L-DOPA in acetic acid to yield a 1:1 mixture of L-2-and L-6-[18F]fluoro-DOPA upon hydrolysis[151] in about 10% radiochemical yield. This method was later preferred for routine synthesis, since it avoided the use of dangerous HF.

Aryl [18F]fluorides can also be prepared by regioselective [18F]fluorodemetallation in metalloarenes, containing Sn, Pb, Hg or Ge. Tetraphenyltin and phenyl tri-n-butyltin react with [^{18}F]fluorine to give [^{18}F]fluorobenzene in only modest yield.[152,153] By contrast p-chloro- and p-fluoro-tri-n-butyltin-arenes react with [^{18}F]fluorine to give substituted aryl [^{18}F]fluorides in very high yields (> 95%).[153,154]

Aryl-silicon bonds in aryl pentafluorosilicates[155,156] and aryltrimethylsilanes[156-159] are also cleaved with [^{18}F]fluorine to give aryl [^{18}F]fluorides, regioselectively. Indeed cleavage at a protected aryltrimethylsilane provides a regioselective synthesis of L-6-[^{18}F]fluoro-DOPA.[160]

Acetyl [^{18}F]hypofluorite also cleaves aryl-metal[153] and aryl-silicon[156] bonds to give aryl [^{18}F]fluorides. A systematic study of the reactions of acetyl [^{18}F]hypofluorite and of [^{18}F]fluorine with aryltrimethyl-tin, -germanium and -silicon compounds found that [^{18}F]fluorination decreased at aryl rings containing electron-withdrawing substituents.[161] The order of reactivity was also found to be aryl-Sn > aryl-Ge > aryl-Si i.e. in inverse order of aryl-metal bond strength. The reactivity of acetyl [^{18}F]hypo-fluorite only matched that of [^{18}F]fluorine with the aryltins. A series of aryl-tri-n-butyltins (R-Ar-SnBu$_3$, R = Cl, F, H, Me, OMe, di-OMe) treated with acetyl [^{18}F]hypo-fluorite gave substituted aryl [^{18}F]fluorides in high radio-chemical yields,[153] comparable to those obtained on these substrates with [^{18}F]fluorine. [^{18}F]Fluorodestannylation is not possible at alkyl-tins, because of the high C-Sn bond strength, but is possible at vinyl-tins. [^{18}F]Fluorodestannylation has been applied to the preparation of p-[^{18}F]fluorophenyl-alanine[162] for the study of protein synthesis. Perhaps the most impressive example of [^{18}F]fluoro-demetallation to date is however a high yield of synthesis of L-6-[^{18}F]fluoro-DOPA by [^{18}F]fluorodemercuration.[163]

At present the use of CA [^{18}F]fluorine and its electrophilic derivatives are the only established methods for introducing ^{18}F into electron-rich aromatic systems. These labelling agents retain their importance mainly on this basis, despite their disadvantages of low specific activity, possibility of only moderately high production and 50% waste of initial activity.

NCA and Nucleophilic. Fluoride ion has a high hydration energy.[164] Thus, in the presence of water, fluoride tends to be highly hydrated and poorly reactive as a nucleophile. The main methods of producing NCA ^{18}F use water as target material (Table 2). Thus the first task, preceding any organic chemistry, is to improve the reactivity of the [^{18}F]fluoride by removing bulk water and water of hydration, in effect to obtain 'naked [^{18}F]fluoride', a powerful nucleophile in polar aprotic media.[165] Removal of water by evaporation concentrates involatile contaminants (e.g. metal ions, nucleophilic anions) that always arise from the target during irradiation. These contaminants may subsequently hinder reactions of NCA [^{18}F]fluoride. Generally their presence is minimised by good target design and is often tolerated in subsequent radiochemistry. It is important to appreciate that much of the work so far reported has been with unpurified [^{18}F]fluoride of high, but often of unknown specific activity. Levels of contaminants and their effects are often unknown. Dilution of [^{18}F]fluoride with carrier increases radiochemical yields in some reactions[166] but conversely decreases radiochemical yields in others.[167] These factors make comparisons of data difficult and sometimes meaningless.

Recently [^{18}F]fluoride has been purified by conversion into [^{18}F]fluorotrimethylsilane with trimethylsilyl chloride, evaporation and hydrolysis.[168,169] Such [^{18}F]fluoride is reported to give reproducible reaction rates in nucleophilic substitutions.[170] [^{18}F]Fluoride can also be purified by distillation from acid solution. Precautions must then be taken against co-distillation of other competitive anions. Ideally it would be convenient to gauge the reactivity of a sample of NCA [^{18}F]fluoride in a broad spectrum of nucleophilic reactions, by measuring the radiochemical yield and rate constant in a single reference reaction. It seems unlikely that this will be possible, since different samples of [^{18}F]fluoride can show different patterns of reactivity.[170]

Several procedures are in common use for the generation of reactive [^{18}F]fluoride from aqueous [^{18}F]fluoride. These share a common strategy : dissolution of the [^{18}F]fluoride with a large counterion in a polar aprotic solvent. Generally water is removed by distillation from an added base (e.g., KOH, K_2CO_3, Rb_2CO_3, $Et_4N^+OH^-$, $nBu_4N^+OH^-$, K^+-18-crown-6 complex, K^+-2,2,2-amino-polyether) and the [^{18}F]fluoride salt dried by azeotropic distillation of added acetonitrile or by heating, sometimes with microwaves. Loss of [^{18}F]fluoride can be substantial if only weak base is present. The residue is then solubilised in solvent containing the substrate for nucleophilic attack. Resolubilisation efficiency depends on the vessel used for evaporation and other factors. A systematic study[171] concluded that a commercially available vessel composed of siliconised glass (a Vacutainer) is preferable to a platinum, borosilicate glass or siliconised borosilicate glass vessel, at least for resolubilisation in tetrahydrofuran. Other workers have used pyrex or glassy carbon satisfactorily. Platinum has given generally better results than glass at the MRC Cyclotron Unit. Resolubilisation can be more difficult if a high degree of dryness has been achieved, especially from glass; though the objective is to generate 'naked' fluoride, inexplicably a number of reactions with [^{18}F]fluoride will tolerate low concentrations of water.[172-174]

A good choice of solvent is important for the success of any reaction with [18F]fluoride. Ideally, the solvent should be stable to base ('naked' fluoride is a strong base), non-nucleophilic (to avoid competitive reactions), polar (for high solvent power), high boiling and thermally stable. Some dipolar aprotic solvents best fit these requirements. Acetonitrile is favoured for aliphatic substitutions.[175] Though DMSO is prone to decompose below its boiling point and can act as an oxidant, it is also useful if high reaction temperatures are required. Other useful solvents are THF, DMF, nitrobenzene, o-dichlorobenzene, dichloromethane and tetramethyl sulphone (sulpholane). Rates of aliphatic substitution with fluoride ion on the macro-scale correlate with the capacity of the solvent-base system to dissolve potassium fluoride.[176,177] For example, the observed rates of fluorination of benzyl bromide at 82°C with potassium fluoride follow the order 18-crown-6-acetonitrile > DMF > 1,2-dimethoxy-ethane > acetonitrile > N-methylpyrrolidone.[176] This effect is not so apparent in aromatic nucleophilic substitutions with fluoride. Here protophilic solvents favour reaction, particularly DMSO and sulpholane.[178] In practice DMSO is probably the most useful solvent for aromatic nucleophilic substitutions with [18F]fluoride. These are generally more difficult than aliphatic substitutions and often require the high temperatures that are possible with DMSO.

Criteria for the choice of added base for nucleophilic substitutions with [18F]fluoride are not yet clearly established. Some workers find tetra-n-butylammonium hydroxide (TBA) suitable for a wide variety of reactions.[174] Others have favoured the use of K^+-2.2.2.-aminopolyether complex and have obtained good yields in aliphatic[179-182] and aromatic nucleophilic substitutions,[183] including the most efficient synthesis of [18F]FDG yet known.[182] TBA can also be used in this synthesis but gives lower radio-chemical yield.

Choice of leaving group is of some importance in aliphatic nucleophilic substitutions with [18F]fluoride. Among the halogens, leaving group ability decreases in the order $I^- > Br^- > Cl^- > F^-$. Iodides are sometimes thermally unstable and so bromides can often be better substrates for reaction. [18F]Fluoride for fluorine is chemically possible, but is only of value if added carrier is acceptable. The same order of leaving group ability has been observed in the gas phase preparation of alkyl [18F]fluorides.[184] High yields can be obtained for substit-utions in alkyl bromides and iodides.[165,179,180,185] Reactions, however, seldom incorporate all radioactivity into product. Sulphonate groups, especially triflyl, mesyl and tosyl, feature prominently in aliphatic nucleophilic substitutions with [18F]fluoride, including useful preparations of [18F]FDG (triflyl displacement)[182,186] and of N-[18F]fluoroalkylspiperones (mesyl displacement).[187] Though triflyl is the most powerful leaving group, in some molecules its presence may lead to elimination rather than to substitution. The mesyl group is then preferred. Though less powerful as a leaving group than either triflyl or mesyl, tosyl still finds some use. An important recent application is the preparation of ω-[18F]fluoroalkyl tosylates by nucleophilic substitution on ω-[18F]alkylditosylates.[188] These compounds have been used as agents for the introduction of [18F]fluoroalkyl groups into organic molecules. For example, spiperone, has been [18F]fluoroalkylated at its amido-N atom.[181] Such [18F]alkylspiperones, especially [18F]ethylspiperone and [18F]propylspiperone, retain the pharmacological profile of the parent compound, but have greater brain uptake, a useful property for PET studies. Cyclic sulphate,[189] sulphite[190] and sulphamate[191,192] are also of occasional value as leaving groups in aliphatic nucleophilic substitutions with [18F]fluoride.

Aromatic nucleophilic substitution with [18F]fluoride, requires activation by the presence of an electron withdrawing substituent (NO_2, CN, COR, CHO) ortho or para to the leaving group, which is generally nitro or halogen. High yields are obtained on simple substrates, but are more difficult to achieve on structures that are larger and sensitive to base. Nonetheless impressive yields have been obtained in the preparation of [18F]spiperone by direct 18F⁻ for nitro group exchange, using K^+.2.2.2-amino-polyether as counterion.[183]

Chemistry with [18F]fluoride has the potential to use high initial activity efficiently and to deliver high specific activity. The rapid development of [18F]fluoride chemistry promises to displace many CA low efficiency routes to important 18F-labelled radiopharmaceuticals. Indeed, this has already happened for the preparation of [18F]FDG,[182,193] and is on the horizon for the preparation of L-6-[18F]fluoro-DOPA.[194]

5. CONCLUSION

For the chemist engaged in the development of radio-pharmaceuticals for PET the cyclotron acts as a source of 'reagents'. From the foregoing discussion three important observations can be made about these 'reagents'. First their quality, in terms of chemical composition, is by normal standards, low, uncertain and variable. Secondly, these 'reagents' have short 'shelf-lives'. Thirdly, they are potentially hazardous to use. These factors pose severe challenges in the elaboration of the chemistry of these radioactive 'reagents' for medical research with PET. Nonetheless, in little over a decade much has been achieved. Several radiopharmaceuticals (e.g. [18F]FDG, [1-11C]-acetate, L-6-[18F]fluoro-DOPA, [11C]methylspiperone and [11C]-methionine) are now established for PET, are in production at several PET centres and are providing valuable information to the medical and scientific community. Much more however remains to be achieved. For example, though radiopharmaceuticals now exist for reasonably comprehensive measurements of the dopamine system in vivo, this is not yet so for many of the other important neurotransmitter systems. Nonetheless, so flourishing is this field of radiochemistry that we can optimistically forecast major advances in this and other areas of application over the next decade.

REFERENCES

1. M.E. Phelps, E.J. Hoffman, N.A. Mullani and M.M. Ter-Pogossian, J. Nucl. Med., 1975, 16 210.
2. C.-Y. Shiue and A.P. Wolf, J. Nucl. Med., 1981, 22, P 58.
3. E. Ehrin, E. Westman, S.-O. Nilsson, J.L.G. Nilsson, L. Widen, T. Greitz, C.-M. Larson, J.-E. Tillberg and P. Malmborg, J. Labelled Compd. Radiopharm., 1980, 17, 453.
4. S.K. Luthra, V.W. Pike, F. Brady, D.R. Turton, B. Wood, R.W. Matthews and G.E. Hawkes,. J. Labelled Compd. Radiopharm., 1986, 23, 1070.
5. M. Reivich, D. Kuhl, A. Wolf, J. Greenberg, M. Phelps, T. Ido, V. Casella, J. Fowler, E. Hoffman, A. Alavi, P. Som and L. Sokoloff, Circulation Res., 1979, 44, 127.
6. M.E. Phelps, S.C. Huang, E.J. Hoffman, C. Selin, L. Sokoloff and D.E. Kuhl, Ann. Neurol., 1979, 6, 371.
7. R.A. Peters, Adv. Enzymol., 1957, 7, 120.
8. T. Fujita, J. Iwasa and C. Hansch, J. Am. Chem. Soc., 1964, 86, 5175.
9. C. Hansch and J. Fukunaga, Chemtech., 1977, 7, 120.
10. C.-Y. Shiue, R.R. MacGregor, R.E. Lade, C.N. Wan and A.P. Wolf, J. Nucl. Med., 1978, 19, 676.
11. G. Kloster, C. Müller-Platz and P. Laufer, J. Labelled Compd. Radiopharm., 1981, 18, 855.
12. T.Z. Csaky and J.E. Wilson, Biochim. Biophys. Acta, 1956, 22, 185.
13. R.F. Dannals, H.T. Ravert, A.L. Wilson and H.N. Wagner Jr., Int. J. Appl. Radiat. Isot., 1986, 36, 433.
14. D.F.Wong, H.N. Wagner Jr., Tune L.E., R.F. Dannals, G.D. Pearlson, J.M. Links, C.A. Tamminga, E.P. Brousolle, H.T. Ravert, A.A. Wilson, J.K.T. Toung, J. Malat, J.A. Williams, L.A. O'Tuama, S.H. Snyder, M.J. Kuhar and A. Gjedda, Science, 1986, 234, 1558.
15. M. Selikson, R.E. Gibson, W.C. Eckelman, R.C. Reba, Anal. Biochem., 1980, 108, 64.
16. G. Stocklin, H.H. Coenen, M.F. Harmand, G. Kloster, E.J. Knust, C. Kupfernagel, H.J. Machulla, R. Weinreich, L.E. Feinendegen, K. Vyska, A. Hock and C. Freundlieb, Radioakt. Isotope Klin. Forsch., 1980, 14, 151.

17. J.S. Fowler and A.P. Wolf, 'Positron Emission Tomography and Autoradiography : Principles and Applications for the Brain and Heart', Eds. M. Phelps, J. Mazziotta and H. Schelbert, Raven Press, New York, 1986, Chapter 9, p. 391.

18. 'Stopping Powers for Electrons and Positrons', International Commission on Radiation Units and Measurements, Report 37, 1984, p. 258.

19. 'Table of Isotopes', Eds. C.M. Lederer and V.S. Shirley, John Wiley and Sons Inc., New York, 7th Edition 1978.

20. E.J. Hoffman and M.E. Phelps, 'Positron Emission Tomography and Autoradiography : Principles and Applications for the Brain and Heart', Eds. M. Phelps, J. Mazziotta and H. Schelbert, Raven Press, New York, 1986, Chapter 6, p. 237.

21. A.P. Wolf, J. Labelled Compd. Radiopharm., 1981, 18, 1.

22. S.M. Qaim and G. Stöcklin, Radiochim. Acta, 1983, 34, 25.

23. G. Stöcklin, Nucl. Med. Biol.,1986, 13, 109.

24. M.J. Welch and K.D. McElvany, Radiochim. Acta, 1983, 34, 41.

25. H.H. Coenen, S.M. Moerlein and G. Stöcklin, Radiochim. Acta, 1983, 34, 47.

26. B. Mazière and C. Loc'h, Appl. Radiat. Isot., 1986, 37, 703.

27. J.C. Baron, B. Maziere, C. Loc'h, P. Sgouropoulos, A.M. Bonnet and Y. Agid, Lancet, 1985i, 1163.

28. H.H. Coenen, M.-F. Harmand, G. Kloster and G. Stocklin, J. Nucl. Med., 1981, 22, 891.

29. A.P. Wolf and W.B. Jones, Radiochim. Acta, 1983, 34, 1.

30. S.M. Qaim, Radiochim. Acta, 1982, 30, 147.

31. 'Proceedings of the First Workshop on Targetry and Target Chemistry', Eds. F. Helus and T.J. Ruth, DKFZ Press Dept. Heidelberg, W. Germany, 1987.

32. R.A. Ferrieri and A.P. Wolf, Radiochim. Acta, 1983, 34, 69.

33. V.R. Casella, D.R. Christman, T. Ido and A.P. Wolf, Radiochim. Acta, 1978, 25, 17.

34. G.T. Bida, T.J. Ruth and A.P. Wolf, Radiochim. Acta, 1980, 27, 181.

35. M. Epherre and C. Seide, Phys. Rev. C, 1971, 3, 2167.

36. D.R. Christman, R.D. Finn, K.I. Karlstrom and A.P. Wolf, Int. J. Appl. Radiat. Isot., 1975, 26, 435.

37. C. Crouzel, B. Längström, V.W. Pike and H.H. Coenen, Appl. Radiat. Isot., 1987, 38, 605.

38. R.S. Tilbury, J.R. Dahl, W.G. Monahan and J.S. Laughlin, Radiochem. Radioanal. Lett., 1971, 8, 317.

39. W. Vaalburg, J.A.A. Kamphuis, H.D. Beerling-van der Molen, S. Reiffers, A. Rijskamp and M.G. Woldring, Int. J. Appl. Radiat. Isot., 1975, 26, 316.

40. T.C. Hollocher, E. Garber, A.J.L. Cooper and R.E. Reiman, J. Biol. Chem., 1980, 255, 5027.

41. V. Casella, T. Ido, A.P. Wolf, J.S. Fowler, R.R. MacGregor and T.J. Ruth, J. Nucl. Med., 1980, 21, 750.

42. G.T. Bida, R.L. Ehrenkaufer, A.P. Wolf, J.S. Fowler, R.R. MacGregor and T.J. Ruth, J. Nucl. Med., 1980, 21, 758.

43. E.J. Knust, H.-J. Machulla and W. Roden, Appl. Radiat. Isot., 1986, 37, 853.
44. M.R. Kilbourn, J.T. Hood and M.J. Welch, Int. J. Appl. Radiat. Isot., 1984, 35, 599.
45. T. Nozaki, Y. Tanaka, A. Shimamura and T. Karasawa, Int. J. Appl. Radiat. Isot., 1968, 19, 27.
46. J. Fitschen, R. Beckmann, U. Holm and H. Neuert, Int. J. Appl. Radiat. Isot., 1977, 28, 781.
47. B.W. Wieland and A.P. Wolf, J. Nucl. Med., 1983, 24, P122.
48. J. Keinonen, A. Fontell and A.-L. Kairento, Appl. Radiat. Isot., 1986, 37, 631.
49. M.B. Winstead, J.F. Lamb and H.S. Winchell, J. Nucl. Med., 1973, 14, 747.
50. V.W. Pike, M.N. Eakins, R.M. Allan and A.P. Selwyn, Int. J. Appl. Radiat. Isot., 1982, 33, 505.
51. V.W. Pike, M.N. Eakins, R.M. Allan and A.P. Selwyn, J. Radioanal. Chem., 1981, 64, 291.
52. H.C. Padgett, G.D. Robinson and J.R. Barrio, Int. J. Appl. Radiat. Isot., 1982, 33, 1471.
53. W. Vaalburg, H.D. Beerling-van der Molen, S. Reiffers, A. Rijskamp, M.G. Woldring and H. Wynberg, Int. J. Appl. Radiat. Isot., 1976, 27, 153.
54. J.M. Bolster, W. Vaalburg, T.H. Van Dijk, J.B. Zijlstra, A.M.J. Paans, H. Wynberg and M.G. Woldring, Int. J. Appl. Radiat. Isot., 1985, 36, 263 and references cited therein.
55. G.A. Digenis, D.L. Casey, D.A. Wesner, L.C. Washburn and R.L. Hayes, J. Nucl. Med., 1979, 20, 662.
56. L.C. Washburn, T.T. Sun, B.L. Byrd and A.P. Callahan, J. Nucl. Med., 1982, 23, 29.
57. J.H.C. Wu, P.V. Harper and K.A. Lathrop, J. Nucl. Med., 1981, 22, P74.
58. S. Ram, R.E. Ehrenkaufer and D.M. Jewett, Appl. Radiat. Isot., 1986, 37, 391.
59. J.M. Bolster, W. Ten Hoeve, W. Vaalburg, T.H. Van Dijk, J.B. Zijlstra, A.M.J. Paans, H. Wynberg and M.G. Woldring, Appl. Radiat. Isot, 1985, 36, 339.
60. J.F. Lifton and M.J. Welch, Radiat. Res., 1971, 45, 35.
61. K. Ishiwata, M. Monma and T. Ido, Appl. Radiat. Isot., 1987, 38, 475.
62. A.J. Palmer and R.W Goulding, J. Labelled Compd. Radiopharm., 1974, 101, 627.
63. C. Marazano, M. Maziere, G. Berger and D. Comar, Int. J. Appl. Radiat. Isot., 1977, 28, 49.
64. B. Längström and H. Lundqvist, Int. J. Appl. Radiat. Isot., 1976, 27, 357.
65. R. Iwata, T. Ido, A. Ujüce, T. Takahashi, K. Ishiwata, K. Hatano and M. Sugahara, Appl. Radiat. Isot., 1988, 39, 1.
66. M.C. Lasne, V.W. Pike and D.R. Turton, Br. J. Radiology, in press.

67. M.R. Kilbourn, D.D. Dischino and M.J. Welch, Int. J. Appl. Radiat. Isot., 1984, 35, 603.

68. B. Längström and B. Stridsberg, Int. J. Appl. Radiat. Isot., 1979, 30, 151.

69. W.K. Dean, K.J. Heald, and S.N. Deming, Science, 1975, 189, 805.

70. W. Spendley, G.R. Hext and F.R.Himsworth, Technometrics, 1962, 4, 441.

71. A. Rimland, G. Bergson, U. Obenius, S. Sjöberg and B. Långstöm, Appl. Radiat. Isot., 1987, 38, 651.

72. G. Berger, M. Maziere, J. Sastre and D. Comar, J. Labelled Compd. Radiopharm., 1980, 17, 59.

73. D.R. Christman, E.J. Crawford, M. Friedkin and A.P. Wolf, Proc. Natl. Acad. Sci. USA., 1971, 69, 988.

74. V.W. Pike, A.J. Palmer, P.L. Horlock, T.J. Perun, L.A. Freiberg, D.A. Dunnigan and R.H. Liss, J. Chem. Soc.,Chem. Commun., 1982, 173.

75. P. Marche, C. Marazono, M. Mazière, J.L. Morgat, P. de la Llosa, D. Comar and P. Fromageot, Radiochem. Radioanal. Lett., 1975, 21, 53.

76. M.M. Vora, R.D. Finn, T.E. Boothe, D.R. Liskowsky and L.T. Potter, J. Labelled Compd. Radiopharm., 1983, 20, 1229.

77. S. Reiffers, W. Vaalburg, T. Wiegman, H. Wynberg and M.G. Woldring, Int. J. Appl. Radiat. Isot., 1980, 31, 535.

78. E. Ehrin, L. Farde, T. De Paulis, L. Eriksson, T. Greitz, P. Johnström, J.-E. Litton, J. L. G. Nilsson, G. Sedvall, S. Stone-Elander and S.-O. Ogren, Int. J. Appl. Radiat. Isot., 1985, 36, 269.

79. G. Slegers, J. Sambre, P. Goethals, C. Vandecasteele and D. Van Haver, Appl. Radiat. Isot., 1986, 37, 279.

80. B. Långström, G. Antoni, P. Gullberg, C. Halldin, K. Någren, A. Rimland and H. Svard, Appl. Radiat. Isot., 1986, 37, 1141.

81. G. Antoni and B.Långström, Appl. Radiat. Isot., 1987, 38, 655.

82. A. Rimland and B. Långstrom, Appl. Radiat. Isot., 1987, 38, 949.

83. G. Antoni and B. Långström, J. Labelled Compd. Radiopharm., 1987, 24, 125.

84. S.K. Luthra, V.W. Pike and F. Brady, J. Chem. Soc., Chem. Commun., 1985, 1423.

85. E. Ehrin, S.K. Luthra, C. Crouzel and V.W. Pike, J. Labelled Compd. Radiopharm., 1986, 23, 1410.

86. D. Le Bars, S.K. Luthra, V.W. Pike and C. Luu Duc, Appl. Radiat. Isot., 1987, 38, 1073.

87. G. Berger, M. Maziere, C. Prenant and D. Comar, Int. J. Appl. Radiat. Isot., 1980, 31, 577.

88. G. Berger, M. Maziere, C. Prenant, J. Sastre, A. Syrota and D. Comar, J. Radioanal. Chem., 1982, 74, 301.

89. G. Berger, C. Prenant, J. Sastre, A. Syrota and D. Comar, Int. J. Appl. Radiat.Isot., 1983, 34, 1556.

90. C. Prenant, J. Sastre, C. Crouzel and A. Syrota, J. Labelled Compd. Radiopharm., 1987, 24, 227.
91. M.J. Welch and M.M. Ter-Pogossian, Radiat. Res., 1968, 36, 580.
92. M.R. Kilbourn, P.A. Jerabek and M.J. Welch, J. Chem. Soc., Chem. Commun., 1983, 861.
93. D.Y. Tang, A. Lipman, G.-J. Meyer, C.-N. Wan and A.P. Wolf, J. Labelled Compd. Radiopharm., 1979, 16, 435.
94. G.A. Brinkman, I. Hass-Lisewska, J. Th. Veenboer and L. Lindner, Int. J. Appl. Radiat., Isot., 1978, 29, 701.
95. D. Roeda, C. Crouzel and B. van Zanten, Radiochem. Radioanal. Lett., 1978, 33, 175.
96. C. Crouzel, G. Mestelan, E. Kraus, J.M. Lecomte and D. Comar, Int. J. Appl. Radiat. Isot, 1980, 31, 545.
97. M. Berridge, D. Comar, C. Crouzel and J.-C. Baron, J. Labelled Compd. Radiopharm., 1983, 20, 73.
98. P. Landais and C. Crouzel, Appl. Radiat. Isot., 1987, 38, 297.
99. C. Crouzel, R. Amano and D. Fournier, Appl. Radiat. Isot., 1987, 38, 669.
100. C. Crouzel, C. Sejourne and D. Comar, Int. J. Appl. Radiat. Isot., 1979, 30, 566.
101. W. Vaalburg, A. Feenstra, T. Wiegman, H.D. Beerling, S. Reiffers, A. Talma, M.G. Woldring and H. Wynberg, J. Labelled Compd. Radiopharm., 1981, 18, 100.
102. W. Vaalburg, S. Reiffers, E. Beerling, J.J. Pratt, M.G. Woldring and H. Wynberg, J. Labelled Compd. Radiopharm., 1977, 13, 200.
103. R. Iwata, T. Ido, T. Takahashi, H. Nakanishi and S. Iida, Appl. Radiat. Isot., 1987, 38, 97.
104. L.C. Washburn, B.W. Wieland, T.T. Sunn, R.L. Hayes and T.A. Butler, J. Nucl. Med., 1978, 19, 77.
105. J. Sambre, C. Vandecasteele, P. Goethals, N.A. Rabi, D. Van Haver and G. Slegers, Int. J. Appl. Radiat. Isot., 1985, 36, 275.
106. R.L. Hayes, L.C. Washburn, B.W. Wieland, T.T. Sun, R.R. Turtle and T.A. Butler, J. Nucl. Med., 1976, 17, 748.
107. D.R. Christman, R.M. Hoyte and A.P. Wolf, J. Nucl. Med., 1970, 11, 474.
108. J.S. Fowler, B.M. Gallagher, R.R. MacGregor and A.P. Wolf, J. Pharmacol. Exp. Ther., 1976, 198, 133.
109 A.J.L. Cooper, A.S. Gelbard and B.R. Freed, Adv. Enzymol., 1985, 57, 251.
110. W.A. Pettit, R.S. Tilbury, G.A. Digenis and R.H. Mortara, J. Labelled Compd. Radiopharm., 1977, 13, 119.
111. G.A. Digenis, Y.C. Chang, R.L. McQuinn, B.R. Freed and R.S. Tilbury in 'Short-Lived Radiopharmaceuticals in Chemistry and Biology', Advances in Chemistry Series 197, J.W. Root and K.A. Krohn, Eds. American Chemical Society, Washington DC, 1981, pp. 351-367.

112. D.S. Kaseman, A.J.L. Cooper, A. Meister, A.S. Gelbard and R.E. Reiman, J. Labelled Compd. Radiopharm., 1984, 21, 803.

113. H. Krizek, N. Lembares, R. Dinwoodie, I. Gloria, K.A. Lathrop and P.V. Harper, J. Nucl. Med., 1973, 14, 629.

114. C.-H. Kim and T.C. Hollocher, J. Biol. Chem., 1983, 258, 4861.

115. A.S. Gelbard, L.P. Clarke and J.S. Laughlin, J. Nucl. Med., 1974, 15, 1223.

116. G. Slegers, C. Vandecasteele and J. Sambre, J. Radioanal. Chem., 1980, 59, 585.

117. T. Tominaga, K. Suzuki, O. Inoue, T. Irie, T. Yamasaki and M. Hirobe, Appl. Radiat. Isot., 1987, 38, 437.

118. M.G. Straatman and M.J. Welch, Radiat. Res.,1973, 56, 48.

119. A.J.L. Cooper and A.S. Gelbard, Anal. Biochem., 1981, 111, 42.

120. J.R. Barrio, F.J. Baumgartner, M.E. Phelps, E. Henze, H.R. Schelbert and N.S. MacDonald, J. Nucl. Med., 1982, 23, P107.

121. F.J. Baumgartner, J.R. Barrio, E. Henze, H.R. Schelbert, N.S. MacDonald, M.E. Phelps and D.E. Kuhl, J. Med. Chem., 1981, 24, 764.

122. M.G. Straatman, Int. J. Appl. Radiat. Isot., 1977, 28, 13.

123. A.J.L. Cooper, J. McDonald, A.S. Gelbard, R.F. Gledhill and T.E. Duffy, J. Biol. Chem., 1979, 254, 4982.

124. A.S. Gelbard and A.J.L. Cooper, J. Labelled Compd. Radiopharm., 1979, 16, 94.

125. M.B. Cohen, L. Spolter, C.C. Chang, N.S. MacDonald, J. Takahashi and D.D. Bobinet, J. Nucl. Med., 1974, 15, 1192.

126. T. Tominaga, O. Inoue, K. Suzuki, T. Yamasaki and M. Hirobe, Appl. Radiat. Isot., 1986, 37, 1209.

127. T. Tominaga, O. Inoue, T. Irie, K. Suzuki, T. Yamasaki and M. Hirobe, Appl. Radiat. Isot., 1985, 36, 555.

128. T. Irie, O. Inoue, K. Suzuki and T. Tominaga, Int. J. Appl. Radiat. Isot., 1985, 36, 345.

129. D.R. Elmaleh, D.J. Hnatowich and S. Kulprathipanja, J. Labelled Compd. Radiopharm., 1979, 16, 92.

130. P.J. Kothari, R.D. Finn, G.W. Kabalka, M.M. Vora, T.E. Boothe and A.M. Emran, Appl. Radiat. Isot., 1986, 37, 469.

131. J. Kollonitsch, in 'Biomedicinal Aspects of Fluorine Chemistry' p.93-122, Eds. R. Filler and Y. Kobayashi, Kodanska Ltd, Tokyo, 1982.

132. M.R. Kilbourn, M.J. Welch, C.S. Dence, T.J. Tewson, H. Saji and M. Maeda, Int. J. Appl. Radiat. Isot., 1984, 35, 591.

133. H.H. Coenen, P. Laufeŋ, G. Stöcklin, K. Wienhard, G. Pawlik, H.G. Böcher-Schwarz and W.-D. Heiss, Life Sciences, 1987, 40, 81.

134. C.-Y. Shiue, P.A. Salvadori, A.P. Wolf, J.S. Fowler and R.R. MacGregor, J. Nucl. Med., 1982, 23, P108.

135. C.-Y. Shiue, P.A. Salvadori, A.P. Wolf, J.S. Fowler and R.R. MacGregor, J. Nucl. Med., 1982, 23, 899.

136. H.C. Padgett, J.S. Cook and J.R. Barrio, <u>J. Nucl. Med.</u>, 1983, <u>24</u>, P121.

137. D.M. Jewett, J.F. Potocki and R.E. Ehrenkaufer, <u>Synth. Commun.</u>, 1984, <u>14</u>, 45.

138. G. Schrobilgen, G. Firmau, R. Chirakal and E.S. Garnett, <u>J. Chem. Soc.,Chem. Commun.</u>, 1981, 198.

139. R.E. Ehrenkaufer and R.R. MacGregor, <u>J. Labelled Compd. Radiopharm.</u>, 1982, <u>19</u>, 1637.

140. S.T. Purrington and B.S. Kagen, <u>Chem. Rev.</u>, 1986, <u>86</u>, 997.

141. T. Ido, C.-N. Wan, V. Casella, J.S. Fowler, A.P. Wolf, M. Reivich and D.E. Kuhl, <u>J. Labelled Compd. Radiopharm.</u>, 1978, <u>14</u>, 175.

142. G.T. Bida, N. Satyamurthy and J.G. Barrio, <u>J. Nucl. Med.</u>, 1984, <u>25</u>, 1327.

143. C.-Y. Shiue, J.S. Fowler, A.P. Wolf, D. Alexoff and R.R. MacGregor, <u>J. Labelled Compd. Radiopharm.</u>, 1985, <u>22</u>, 503.

144. G.T. Bida, N. Satyamurthy, H.C. Padgett and J.R. Barrio, <u>J. Labelled Compd. Radiopharm.</u>, 1984, <u>21</u>, 1196.

145. C.J.S. van Rijn, J.D.M. Herschied, G.W.M. Visser and A. Hoekstra, <u>Int. J. Appl. Radiat. Isot.</u>, 1985, <u>36</u>, 111.

146. R.E. Ehrenkaufer, J.F. Potocki and D.M. Jewett, <u>J. Nucl. Med.</u>, 1984, <u>25</u>, 333.

147. C.-Y. Shiue, R.S. Kutzman and A.P. Wolf, <u>Eur. J. Nucl. Med.</u>, 1985, <u>10</u>, 278.

148. M. Diksic, S. Farrokhzad, L.Y. Yamamoto and W. Feindel, <u>Int. J. Nucl. Med. Biol.</u>, 1984, <u>11</u>, 141.

149. C.-Y. Shiue, A.P. Wolf and M. Friedkin, <u>J. Labelled Compd. Radiopharm.</u>, 1982, <u>19</u>, 1395.

150. G. Firnau, R. Chirakal and E.S. Garnett, <u>J. Nucl. Med.</u>, 1984, <u>25</u>, 1228.

151. M.J. Adam, T.J. Ruth, J.R. Grierson, B. Abeysekera and B.D. Pate, <u>J. Nucl. Med.</u>, 1986, <u>27</u>, 1462.

152. M.J. Adam, B.D. Pate, T.J. Ruth, J.M. Berry and L.D. Hall, <u>J. Chem. Soc.,Chem. Commun.</u>, 1981, 733.

153. M.J. Adam, T.J. Ruth, S. Jivan and B.D. Pate, <u>J. Fluorine Chem.</u>, 1984, <u>25</u>, 329.

154. M.J. Adam, B.F. Abeysekera, T.J. Ruth, S. Jivan and B.D. Pate, <u>J. Labelled Compd. Radiopharm</u>, 1984, <u>21</u>, 1227.

155. M. Speranza, C.-Y. Shiue, A.P. Wolf, D.S. Wilbur and G. Angelini, <u>J. Chem. Soc.,Chem. Commun.</u>, 1984, 1448.

156. M. Speranza, C.-Y. Shiue, A.P. Wolf, D.S. Wilbur and G. Angelini, <u>J. Fluorine Chem.</u>, 1985, <u>30</u>, 97.

157. M.J. Adam, J.M. Berry, L.D. Hall, B.D. Pate and T.J. Ruth, <u>Can. J. Chem.</u>, 1983, <u>61</u>, 658.

158. P. DiRaddo, M. Diksic and D. Jolly, <u>J. Chem. Soc., Chem. Commun.</u>, 1984, 159.

159. M. Diksic, S. Farrokhzad and P. DiRaddo, <u>J. Labelled Compd. Radiopharm.</u>, 1984, <u>21</u>, 1187.

160. M. Diksic and S. Farrokhzad, <u>J. Nucl. Med.</u>, 1985, <u>26</u>, 1314.

161. H.H. Coenen and S.M. Moerlein, J. Fluorine Chem., 1987, 36, 63.
162. H.H. Coenen, K. Franken, S. Metwally and G. Stöcklin, J. Labelled Compd. Radiopharm., 1986, 23, 1179.
163. A. Luxen, J.R. Barrio, G.T. Bida and N. Satyamurthy, J. Labelled Compd. Radiopharm., 1986, 23, 1066.
164. W.A. Sheppard and C.M. Sharts, Organic Fluorine Chemistry, 1969, W.A. Benjamin, New York.
165. S.J. Gatley and W.J. Shaughnessy, Int. J. Appl. Radiat. Isot., 1982, 33, 1325.
166. E.J. Knust, M. Schüller and G. Stöcklin, J. Labelled Compd. Radiopharm., 1980, 17, 353.
167. K. Hamacher, H.H. Coenen and G. Stöcklin, J. Labelled Compd. Radiopharm., 1986, 23, 1095.
168. M.S. Rosenthal, A.L. Bosch, R.J. Nickles and S.J. Gatley, Int. J. Appl. Radiat. Isot., 1985, 36, 318.
169. L.G. Hutchins, A.L. Bosch, M.S. Rosenthal, R.J. Nickles and S.J. Gatley, Int. J. Appl. Radiat. Isot., 1985, 36, 375.
170. R.J. Nickles, S.J. Gatley, J.R. Votaw and M.L. Kornguth, Appl. Radiat. Isot., 1986, 37, 649.
171. J.W. Brodack, M.R. Kilbourn, M.J. Welch and J.A. Katzenellenbogen, Appl. Radiat. Isot., 1986, 37, 217.
172. J.R. Ballinger, B.M. Bowen, G. Firnau, E.S. Garnett and F.W. Teare, Int. J. Appl. Radiat. Isot., 1984, 35, 1125.
173. F. Cacace, M. Speranza and A.P. Wolf, J. Fluorine Chem., 1982, 21, 145.
174. M.R. Kilbourn, J.W. Brodack, D.Y. Chi, C.S. Dence, P.A. Jerabek, J.A. Katzenellenbogen, T.B. Patrick and M.J. Welch, J. Labelled Compd. Radiopharm., 1986, 23, 1174.
175. S.J. Gatley, Int. J. Appl. Radiat. Isot., 1982, 33, 255.
176. C.L. Liotta and H.P. Harris, J. Am. Chem. Soc., 1974, 96, 2250.
177. T. Kitazume and N. Ishikawa, Chem. Lett., 1978, 283.
178. J.H. Clark and D. Macquarrie, J. Fluorine Chem., 1987, 35, 591.
179. H.H. Coenen, B. Klatte, A. Knöchel, M. Schüller and G. Stöcklin, J. Labelled Compd Radiopharm., 1986, 23, 455.
180. H.H. Coenen, M. Colosimo, M. Schüller and G. Stöcklin, J. Labelled Compd. Radiopharm., 1986, 23, 587.
181. H.H. Coenen, P. Laufer, G. Stöcklin, K. Wienhard, G. Pawlik, H.G. Böcher-Schwarz and W.-D. Heiss, Life Sciences, 1987, 40, 81.
182. K. Hamacher, H.H. Coenen and G. Stöcklin, J. Nucl. Med., 1986, 27, 235.
183. K. Hamacher, H.H. Coenen and G. Stöcklin, J. Labelled Compd. Radiopharm., 1986, 23, 1047.
184. M. Yagi, Y. Murano and G. Izawa, Int. J. Appl. Radiat. Isot., 1982, 33, 1335.
185. H.H. Coenen, M. Schüller and G. Stöcklin, J. Labelled Compd. Radiopharm., 1984, 21, 1197.

186. S. Levy, D.R. Elmaleh and E. Livni, J. Nucl. Med., 1982, 23, 918.
187. D.O. Kiesewetter, R.D. Finn and W.C. Eckelman, J. Labelled Compd. Radiopharm., 1986, 23, 1040.
188. D. Block, H.H. Coenen, P. Laufer and G. Stöcklin, J. Labelled Compd. Radiopharm., 1986, 23, 1042.
189. T.J. Tewson, J. Nucl. Med., 1983, 24, 718.
190. T.J. Tewson and M.J. Welch, J. Nucl. Med., 1980, 21, 559.
191. T.A. Lyle, C.A. Magill and S.M. Pitzenberger, J. Am. Chem. Soc., 1987, 109, 7890.
192. S.K. Luthra, F. Brady and V.W. Pike, Br. J. Radiology. In press.
193. H.H. Coenen, V.W. Pike, G. Stöcklin and R. Wagner, Appl. Radiat. Isot., 1987, 38, 605.
194. C. Lemaire, M. Guillaume, L. Christiaens, A.J. Palmer and R. Cantineau, Appl. Radiat. Isot., 1987, 38, 1033.

2
Radioiodination Techniques

D. Silver

AMERSHAM INTERNATIONAL PLC, FOREST FARM, WHITCHURCH, CARDIFF CF4 7YT, UK

1. APPLICATIONS OF RADIOIODINATED LIGANDS

Radioiodinated compounds have a wide range of applications. Their major use is in immunoassay, both radioimmunoassay (RIA), and immunoradiometric assay. The RIA system depends on competition between radio-labelled and unlabelled antigen for a fixed, limited amount of antibody. The labelled and unlabelled materials need not be identical as long as both are similarly recognised by the antibody binding site. Because many biologically active materials are only present in picomole or smaller quantities, very sensitive assays are required. To achieve this, very small amounts of high specific activity tracers are needed. $[^{125}I]$Iodine is particularly suitable for this purpose, being available at a specific activity of ~2000 Ci/mmol. Being a γ-emitting isotope, counting is also easy. A variation on this technique is the competitive protein binding assay. A binding protein, eg cortisol binding globulin, is used rather than an antibody. In immunoradiometric assays (IRMA) the antibody is radiolabelled rather than the analyte to be measured. Excess antibody is used. A second, solid phase coupled, antibody to the antigen is added to facilitate separation of free from bound radio-labelled antibody.

Radioiodinated ligands are used extensively in the study of receptors. Radioreceptor assays are used to measure amounts of receptor in a manner similar to RIA: a receptor preparation is used in place of an antibody. The radiolabelled ligands may also be used to locate the receptor. After allowing the iodinated ligand to bind to the receptor, its distribution can be detected using autoradiography. Examples of this type of radioiodinated ligand are (-)3-[^{125}I]iodo-cyanopindolol, which is used as a probe for the ß-adrenergic receptor and [^{125}I]iodo-α-bungarotoxin, which is used to quantify the nicotinic acetylcholine receptor. The structural requirement of the iodinated ligand can be greater for receptor studies than for RIA, with any deviation from the unlabelled structure having a far greater effect.

Iodinated second antibodies and iodinated protein A or protein G can offer advantages over other detection systems in immunological applications. They are simple to use, can be easily quantified in sensitive assays and can be counted directly, cf enzymes where a substrate is needed. They can be used to screen for positive monoclonal antibody clones, and are used in immunoblotting. Antibodies are particularly useful as a good [^{125}I]-labelled second antibody allows detection of all Ig classes. [^{125}I]Protein A is a general purpose reagent which will bind to many classes of Ig of a variety of species. [^{125}I] Protein G recognises more species and subclasses than [^{125}I]protein A.

Labelled monoclonal antibodies, particularly those labelled with [^{131}I]iodine, have become widely used with the upsurge in the use of monoclonal antibodies in cancer studies. They can be used to localize both tumours and normal cells. I-131 can be used in therapy where ß-radiation from the isotope is used to destroy malignant cells.

Iodine isotopes, both I-125 and I-131, have been used in 'in vivo' studies for a considerable time. They are used in localization studies, eg [^{125}I] iodofibrinogen to detect blood clots, and in function studies, eg iodohippuran which is used to measure kidney function.

2. CHOICE OF TECHNIQUE

There is a range of techniques available for labelling molecules both radioactively and non-radioactively. Of the non-radioactive methods the most widely used to date are probably enzyme labelling and labelling with fluorescent dyes such as fluorescein. Fluorescent dyes have been used in histological studies for many years to detect both antigens and antibodies. Labelled materials can be seen under a fluorescence microscope or measured in a fluorimeter. Enzyme labelled materials are now commonly used. Many are based on the enzyme horseradish peroxidase, though ß-galactosidase and alkaline phosphatase are also often used. The labelled products of their reactions are detected using colour reactions with substrates. Chemiluminescence and enhanced luminescence techniques are also now available.

One of the advantages of these techniques over radioactive methods is the obvious hazard of handling radioactivity. Some non-isotopic techniques also do not require separation of bound from unbound material, unlike RIA. The advantage of isotopic techniques, especially when using iodine isotopes, is the ease of detection of signal. This encompasses the convenience of detection, i.e. direct counting for gamma emitting isotopes, the precise end point of reaction, (cf enzymes and colour formation), and the reproducibility of the system (endogenous enzymes or fluorescence quenchers can occur in biological systems). High specific activity tracers can provide sensitivity in assays which cannot always be matched by non-radioactive techniques.

Choice of Isotope

The most commonly used radioactive isotopes for 'in vitro' applications are H-3, C-14, P-32, S-35, I-125 and I-131. Unlike the other isotopes, in most cases where iodine is used to label a molecule it is a foreign label, i.e it does not normally occur in the molecule. This is not always the case eg thyroxine, triiodothyronine. The replacement of non-radioactive carbon or hydrogen by $[^{14}C]$carbon or tritium will have virtually no effect on the biological properties of the molecule. The replacement of a proton with a large iodine atom can have a considerable effect on the properties of a small molecule and can also affect large molecules. In peptides and proteins this can usually be overcome if the material is labelled at a point distant from the site of biological activity.

There are several major advantages in using $[^{125}I]$iodine over $[^{14}C]$carbon or tritium. The first is in the specific activity available (see Table 1) :

Table 1 : Half life and available specific activity of
commonly used isotopes

Isotope	Half Life	Specific Activity /m atom
I-125	60d	2000 Ci
I-131	8d	2600 Ci
C-14	5600y	62 mCi
H-3	12.2y	29 Ci
P-32	14d	6000 Ci
S-35	87d	1500 Ci

There is an inverse relationship between the half
life of an isotope and its maximum theoretical
specific activity. In some isotopes this maximum is
never obtainable. [^{125}I]Iodine has a maximum
theoretical specific activity of 2200 Ci/mA and is
usually obtained at ~2000 Ci/mA. The maximum specific
activities of [^{14}C]carbon and tritium are 62mCi/mA and
29 Ci/mA respectively. Several atoms of [^{14}C]carbon
or tritium can be substituted in a molecule, but the
specific activity obtained is still very much lower
than with [^{125}I]iodine. Very small amounts of radio-
iodinated materials can be used to give very high
count rates and can be used in sensitive assays. The
count rate obtained from [^{125}I]iodine can be 100 times
greater than for tritium and 35,000 times greater than
[^{14}C]carbon (8).

Another major advantage is in the ease of
detection. [^{125}I]Iodine is a gamma emitter and can be
counted directly in a gamma counter. Both [^{14}C]carbon
and tritium are beta emitters. To count these,
scintillants and a scintillation counter are required.
This involves extra time in sample preparation, extra
counting time, extra cost in scintillant and increased
volumes of radioactive material for disposal.

The high specific activity and count rate of
iodinated compounds is an advantage in autoradiography, especially when very small amounts of receptor
are to be localized. The time required to autoradiograph tritiated and [^{14}C]labelled ligands can stretch
to months.

Complex organic chemistry may be required to
label a molecule with [^{14}C]carbon. This can mean
starting from [^{14}C]-labelled CO_2, methanol, Ba_2CO_3,

benzene etc. It is also an expensive isotope to obtain, and multi-stage preparations inevitably decrease overall yields. Tritiation of samples often involves catalytic hydrogenation to add to a double bond or to replace a halogen in a molecule. Radioiodinations are comparatively easy.

There are also disadvantages in using iodine. As previously stated, iodine is usually a foreign label, and labelling with [^{125}I]iodine can therefore alter the properties of many molecules. This can be a particular problem in receptor studies. Even a small change in structure, such as oxidation of one amino acid in an iodination, can completely block binding to the receptor. Reaction rates can also be altered. The advantage of having a high specific activity is countered by the disadvantage of a shorter half-life. There is also the possibility of faster decomposition, especially radiation decomposition, and also loss of iodine. [^{14}C]Labelled materials can remain pure for many years.

[^{131}I]Iodine is not often used in assays. It has a very short half-life and decays with both β and γ emissions. The counting efficiency is also less than for [^{125}I]iodine. The maximum theoretical specific activity is ~16,000 Ci/mA but it is usually obtainable at only <20% of this. [^{35}S]Sulphur and [^{32}P]phosphorus are usually only used to replace the naturally occurring isotopes.

3.   IODINATION METHODS

There are a variety of methods available for labelling molecules with iodine which lead to the production of materials with high biological activity. These can be divided into direct and indirect iodination. In direct methods iodine is incorporated into a tyrosine or occasionally a histidine residue, which is either part of the molecule to be labelled (often a peptide or protein) or has previously been attached to it. Indirect iodination involves the conjugation of a pre-iodinated species, usually via an amino function on the molecule to be labelled. An important feature required for both types of iodination is the purity of the material to be labelled. Starting with a pure material makes purification of the iodinated products much easier: it is possible that contaminating materials can be more easily iodinated than the ligand itself. Direct iodinations are easier to carry out and tend to give a higher degree of incorporation than indirect methods.

Chloramine T Method

This is the method most commonly used to radio-iodinate peptides and proteins to high specific activity. Chloramine T is the sodium salt of the N-monochloro derivative of p-toluenesulphonamide. This breaks down slowly in aqueous solution, producing hypochlorous acid. This oxidises sodium [^{125}I]iodide to [^{125}I$^+$] which is incorporated into aromatic rings. Excess chloramine T and free iodine are then both reduced. Carrier KI is often added, as is a protein-containing buffer to prevent losses of labelled material during purification. The degree of iodination is concentration dependent, and the reaction needs to be carried out at high ligand concentration.

Oxidation damage can occur during the reaction, especially to methionine residues which can be converted to the sulphoxide. The minimum amount of chloramine T required to produce the degree of iodination and specific activity required should be used.

Commercially obtainable sodium [^{125}I]iodide is supplied in sodium hydroxide solution at pH 8-10. The optimum pH for the iodination of tyrosine is pH 7.2 - 7.5 with reduced incorporation being obtained with pH <6.5 or >8.5. It is, therefore, necessary to buffer the reaction to ~pH 7.2. To iodinate the imidazole ring of histidine a pH of 8.1 is required.

To reduce oxidation damage the reaction time should be as short as possible. This can be seconds for simple compounds, but can be longer for compli-cated peptides with more restricted iodination sites. Another possible problem is the formation of di-iodo material, which appears to be time-dependent. To reduce these problems peptide or protein solutions should be rapidly added and mixed.

The final stage of the reaction is the addition of reducing agent. Sodium metabisulphite was the agent originally employed, but this itself can damage proteins. Cysteine, methionine or tyrosine are now commonly used. Cysteine, however, should not be used if the material being radiolabelled itself contains disulphide bridges. Carrier KI and a protein-containing buffer can then be added if desired, depending on the purification method chosen.

Other Oxidising Agents

To try to overcome the problem of oxidation by chloramine T, other oxidising agents have been used. These include sodium hypochlorite, chlorine and N-bromosuccinimide.

Iodogen

One method of avoiding oxidation problems is to use a water insoluble agent. This avoids direct contact between the oxidising agent and the material to be labelled. 1,3,4,6-tetrachloro-3α,6α-diphenyl-glycoluril (Iodogen) is commonly used. This material is dissolved in dichloromethane and then coated onto the walls of the reaction vessel by evaporating the solvent in a stream of nitrogen. Buffer, sodium [^{125}I]iodide and peptide/protein are added and the solution mixed. Because of the heterogeneous nature of the reaction, longer reaction times are needed. The reaction is terminated by removal of the reaction mixture from the vessel or by addition of a reducing agent. Materials being labelled by this method have to be kept in carrier-free solution (protein free) for a considerable period of time, which can be a problem with proteins.

Electrolytic Method

Electrolysis has been used to convert iodide to iodine with no oxidising agent. This method suffers from the same disadvantage as the iodogen method with even longer reaction times. Special cells are also required.

Enzymatic Method

A widely used method utilizes the enzyme lacto-peroxidase. In the presence of small amounts of hydrogen peroxide, this enzyme oxidises [^{125}I] to [^{125}I$^+$]. Reaction is terminated by dilution with buffer or by quenching the enzyme action with cysteine. Reaction times are long, ~30 mins, but as no strong oxidising or reducing agents are used immunological damage is kept to a minimum. If the presence of small amounts of hydrogen peroxide is a problem this can be produced 'in situ' using glucose/glucose oxidase. The enzyme itself is iodinated using this method, but can usually be removed by HPLC purification or, if gel filtration is to be used, a solid phase coupled enzyme can be employed. This method can give low yields and incorporation at low specific activities.

Iodine monochloride

This was the original iodination method. Iodine monochloride is produced unlabelled and is exchanged with labelled iodine. Inactive iodine is always incorporated by this method, which is not often used.

Bolton and Hunter Reagent

This is by far the most widely used indirect method. Bolton and Hunter reagent is N-succinimidyl 3-(4-hydroxy-5-[^{125}I]iodophenyl)propionate, an active ester. The reagent is itself prepared by a chloramine T iodination (see Fig. 1). This will react with free amino groups of peptide and proteins to form amides. The reagent is commercially available in benzene solution and this is removed by evaporation before reaction.

Conjugation is carried out in buffer around pH 8.5, often at ice bath temperatures. The reaction is concentration dependent, and the reagent itself is hydrolysed slowly to 3-(4-hydroxy 5-[^{125}I]iodophenyl) propionic acid in water. Tyrosine or glycine are added after about 30 mins to conjugate any remaining reagent. This method overcomes any problems of contact with oxidising or reducing agents, and can be used when tyrosine residues occur in the biologically active region of a protein or peptide. It can only be used when a free amino group on a lysine residue is present, or to label a free amine at the N-terminus. Material is commercially available at ~2000 Ci/mmol, but the activity of the product can be increased to 4000 Ci/mmol by using the diiodo material.

Side products from the reaction are of low molecular weight and can be removed by gel filtration or HPLC. As the reagent is itself hydrolysed in buffer, incorporation of iodine tends to be lower than for direct methods though larger amounts of reagent can be used.

Figure 1 : Bolton and Hunter reagent

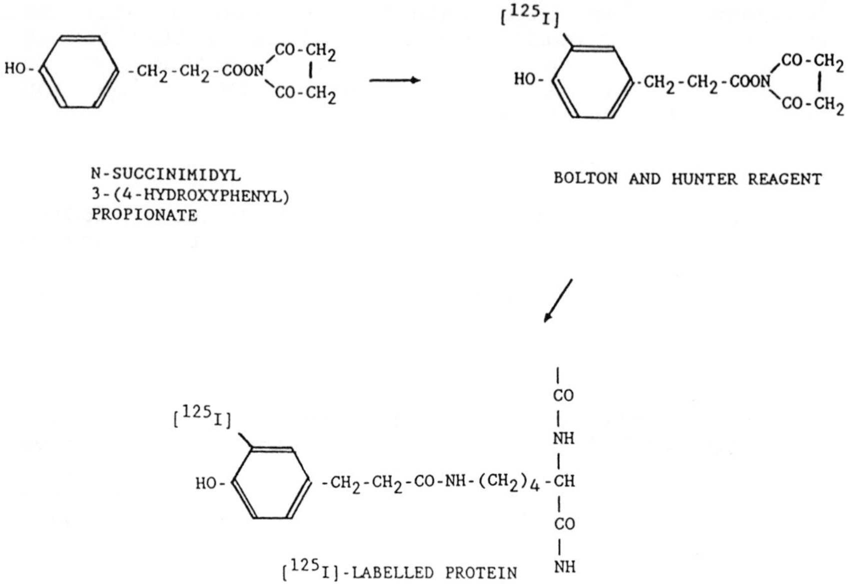

N-SUCCINIMIDYL
3-(4-HYDROXYPHENYL)
PROPIONATE

BOLTON AND HUNTER REAGENT

[^{125}I]-LABELLED PROTEIN

Other Conjugating Agents

Many other conjugating agents have been used, including pre-iodinated tyramine, histamine, tyrosine methyl ester, p-hydroxybenzimidate and fluorescein isothiocyanate. Some of these are particularly useful for small molecules with no amino function or tyrosine. Iodohistamine or iodotyrosine are commonly conjugated to a steroid carboxymethyloxime derivative. For small molecules TLC or HPLC purification is usually essential.

4.   MEASUREMENT OF SPECIFIC ACTIVITY

To measure specific activity of a radioiodination product, a system which will separate iodide from iodinated material is required. This allows calculation of the amount of iodine incorporated. The most commonly used methods are gel filtration and HPLC though electrophoresis and TLC are also used.

Electrophoresis or TLC of an aliquot of the reaction mixture will, by scanning the paper or plate, give a direct value of percentage incorporation of iodine. As a known amount of [^{125}I]iodine is used in the reaction, and a known weight of ligand is iodinated, the specific activity can be calculated.

Gel filtration is particularly useful. Many peptides and proteins are notoriously 'sticky' and adhere strongly to glass. Most iodinations are consequently carried out in plastic vials. Using gel filtration the amount of iodine <u>not</u> incorporated can be measured directly after separation on a column. The amount actually incorporated is, therefore, known by subtraction. The specific activity can be calculated as

$$\frac{\text{Amt } [^{125}\text{I}] \text{ used in reaction} - \text{Amt not incorporated}}{\text{Weight of ligand used}}$$

This method however does not usually give the specific activity of each individual reaction product.

HPLC allows separation of all reaction products. Calculations are not required if HPLC is used to remove all unreacted inactive material. The specific activity of the product will be exactly the same as that of the [^{125}I]iodide or Bolton and Hunter reagent used for reaction. This method is commonly used for small molecules and peptides where iodination increases hydrophobicity and reverse phase HPLC can be used.

5.   PURIFICATION OF IODINATED LIGANDS

Following an iodination, the reaction mixture will contain a mixture of products which will depend on the method of iodination. These include low molecular weight products, eg excess chloramine T, sodium iodide, reducing agents, iodohistamine. High molecular weight materials may also be present, eg lactoperoxidase. Side reactions will also produce a range of products which can include aggregates of peptides or proteins and di-iodinated materials.

In the case of a peptide containing a methionine residue, oxidised peptide will also be obtained if an oxidative method has been used for labelling. Unlabelled ligand, if present, will lower the specific activity of the product, decreasing assay sensitivity. Other impurities may give rise to high non-specific binding. Multi-iodinated species, eg di- and tri-iodinated peptides if present can lead to impaired stability of the product.

Using an example of a model peptide containing two tyrosines and one methionine, on oxidative iodination a total of 17 different products can be formed (1). These include various combinations of multi-iodinated species and mono-iodinated species, both oxidised and unoxidised. This structure occurs in many different peptides including glucagon and VIP.

Many different methods have been used to purify iodinated ligands. These include :

        Gel filtration
        Electrophoresis
        Thin layer Chromatography
        Ion exchange Chromatography
        High performance liquid chromatography
        Affinity Chromatography

Affinity chromatography is particularly useful when labelling immunologically active materials and receptor ligands. However, the conditions required to remove the iodinated ligand from the affinity material are usually fairly harsh and can damage the product. A good example of the use of this method is the purification of [^{125}I]protein A. This material is widely used as a detection reagent for antibodies in screening and in blotting applications where very low backgrounds are required.

After initial iodination using Bolton and Hunter reagent, excess reactants and hydrolysed reagent can be removed by gel filtration. The Protein A can then be further purified using an affinity column of Sepharose-Rabbit IgG. The material is loaded on the column at neutral pH and impurities washed off. The product is then removed by elution with a low pH buffer. Carefully controlling the system allows removal of the material with no apparent damage to the protein, as measured by its binding to various animal immunoglobulin classes. Affinity chromatography does not, however, remove unlabelled material.

The methods of purification most widely used at present are gel filtration, ion exchange chromatography and HPLC (see Table 2). Gel filtration (eg G25 Sephadex) is commonly used to remove low molecular weight impurities such as chloramine T, excess sodium iodide and reducing agents, after iodination and especially after protein iodinations. In many cases this purification will be adequate for subsequent use of the material, especially if only a low specific activity material is required. Gel filtration will remove low molecular weight impurities but high molecular weight material such as diiodinated peptides, aggregates and oxidised material will co-chromatograph with mono-iodinated material. Ion exchange chromatography, which separates by charge, can be used to remove di-iodinated products and unlabelled material as well as lactoperoxidase, but will not separate various multi-iodinated species or oxidised material. HPLC can be used to separate various mono-iodinated materials from all other reaction products and yield a pure product at maximum specific activity.

An example of the advantage of HPLC can be seen from some work carried out on the peptide glucagon (1). Glucagon contains two tyrosine residues at position 10 and 13, and a methionine at position 27. Using a C-terminally directed rabbit antiserum, [^{125}I] glucagon purified by ion exchange chromatography gave a poor standard curve in RIA. This was probably due to oxidation of the methionine.

Two mono-iodinated tracers prepared by reverse phase HPLC were almost indistinguishable and gave good standard curves. These contain intact methionines.

6.   POSITION OF LABELLING

A number of different monoiodinated products can be obtained when a peptide containing more than one tyrosine is iodinated. One of these will be the preferred tracer for RIA or radioreceptor assay. HPLC can be used to separate various mono-iodinated species and produce radiochemically pure high specific activity tracers.

Table 2 : Purification of I-125 labelled peptides

	Gel Filtration	Ion Exchange	RP-HPLC
I-125 iodide	+	+	+
Oxidised peptide	-	-	+
Unreacted peptide	-	+	+
Di- and multi-iodinated peptide	-	+	+
Aggregated peptide	-	?	+
Lactoperoxidase	-	+	+
Monoiodinated isomers	-	-	+

Insulin contains tyrosine residues at positions 14 and 19 in the A chain and at positions 16 and 26 in the B chain. On iodination all four possible mono-iodinated products can be obtained. These can be separated by HPLC (Fig 2) based on the method of Frank et al (2,3) using an ammonium acetate/acetonitrile mixture as the mobile phase. The (A14) and (B26) isomers have been tested and shown to have different properties.

The (A14) isomer behaves similarly to naturally occurring insulin in receptor studies with isolated adipocytes, IM-9 lymphocytes, cultured fibroblasts and isolated hepatocytes (3,4). The (B26) isomer shows increased binding in some tissues and an increase in biological activity indicating that it is a 'super agonist' of native insulin. This has been shown in extra-hepatic tissues such as cultured human fibroblasts, IM-9 lymphocytes and human placental membrane (2,5). In hepatic tissues the (B26) isomer exhibits a lower binding than (A14) [^{125}I]iodoinsulin. When compared in RIA (1) using a guinea-pig antiserum the (A14) isomer gave a higher zero binding and behaved similarly to natural insulin.

In studies using conventionally prepared 'monoiodinated' glucagon, it was postulated that there were two populations of glucagon receptors, one coupled to adenylate cyclase, the other uncoupled (6). Subsequent work using the defined monoiodinated product [^{125}I]-tyr^{10} glucagon suggested that, in fact, all receptors are coupled to adenylate cyclase and that the earlier erroneous conclusions were due to the differing rates of binding and release of glucagon, iodinated at different positions (7).

Fig.2   Preparation of [^{125}I](A14) monoiodoinsulin by reverse phase HPLC – Solvent A, 0·2M ammonium acetate, pH5·5. Solvent B, acetonitrile.

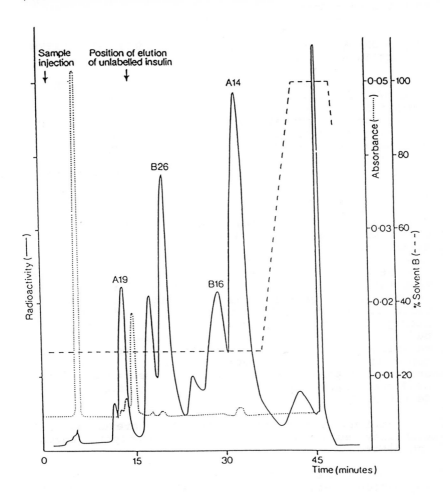

A further advantage of using a mono-iodinated product in a defined position is the removal of batch to batch variation.

7.   STORAGE   AND   STABILITY   OF   RADIO-IODINATED MATERIALS

Many factors are involved in limiting the shelf life of radio-iodinated materials (8). These include radiation decomposition, loss of iodide, aggregation of peptides and proteins and loss of immunoreactivity. Radiation decomposition can be minimised by reducing the degree of di-iodination. A good example is the case of [125I]labelled antibodies where a ratio of one iodine: one antibody molecule seems to confer reasonable stability. Lowering the level of iodination further increases stability in many cases, but is often not practical. Loss of iodine is also most prevalent from multi-iodinated species. Aggregates which may occur on storage can be removed by gel filtration.

Stability on storage can be studied at various temperatures. An example of stability data for [125I] iodohistidyl[15] gastrin releasing peptide is given in Table 3. Lyophilisation is the preferred method of storage for most iodinated peptides. Once dissolved in buffer, storage at -20°C is required. For some materials -70°C storage is better. For most peptides rapid freezing of aliquoted samples is recommended, and freeze-thaw cycles are not desirable. Some products though can be deactivated by freezing, eg ferritin.

Various agents can be added to increase general stability such as aprotinin, a proteolytic enzyme inhibitor.

The choice of the material to label and the iodination method used can also be important. [125I](tyr[11]) somatostatin has been shown to be more stable than the tyr[1] and tyr[0] analogues (9,10). Labelling of CCK-8(SO_4) with Bolton and Hunter reagent not only leaves the residue required for receptor binding in its natural form, but gives a tracer more stable than that produced by the direct labelling method (11).

Table 3 : Stability of [^{125}I]GRP on Storage

Initial RCP=96%	RCP % 5 weeks	RCP% 9 weeks
Lyophilised 4°C	93	88
Aqueous Solution 4°C	25	24
Aqueous Solution -20°C	88	87

RCP = Radiochemical purity

REFERENCES

1. K McFarthing et al, Bioactive analytes including CNS drugs, Peptides and Enantiomers, Plenum Pub Corp, 1986, 65

2. B H Frank et al, Diabetes, 1983, 32, 705

3. B H Frank et al, J.Chromatography, 1983, 266, 239

4. S Linde and B Hansen, Int.J.Peptide and Protein Res, 1980, 15, 495

5. D A Podlecki et al, Diabetes, 1983, 32, 697

6. L Birnbaumer and S L Pohl, J.Biol.Chem., 1973, 248, 2056

7. F J Rojas et al Endocrinology, 1983, 113, 711

8. A E Bolton "Radioiodination techniques", Review 18, Amersham International plc

9. C B Srikant and Y C Patel, Endocrinology, 1982, 110, 2138

10. J M Conlon et al, Biochem.Biophys.Acta,, 1981, 677, 234

11. L J Miller et al, J.Biol.Chem., 1981, 256 12417

3
The Radiochromatography of Labelled Compounds

W.J.S. Lockley

DEPARTMENT OF DRUG METABOLISM, FISONS PLC, PHARMACEUTICAL DIVISION, LOUGHBOROUGH,
LE11 0RH, UK

1. INTRODUCTION

In the ten years since the publication of the last excellent and
comprehensive review of the subject[1], the use of radiochromatography
has spread to a wider range of disciplines. Indeed, so large is the
extent of the literature which describes the use of
radiochromatographic techniques, especially in the medical,
paramedical and life sciences areas, that an exhaustive review would
be undesirable. It would more describe the progress made in these
areas than provide information on radiochromatographic techniques per
se. In reviewing the literature I have consequently restricted the
scope of the task to publications describing substantive points of
radiochromatographic technique rather than the routine applications
where the use of radiochromatography was incidental to the main
purpose of the work. That said, I have nevertheless included some
references of the latter type when I believe that these contribute
new ideas or exemplary applications of techniques. Furthermore, since
the aim of this chapter is to act as an introductory survey of the
subject, I have limited the detail in which each aspect is covered,
trusting that sufficient source information has been included to form
the basis of a more detailed literature search by those with a
specific interest in particular techniques or areas. In this regard
since the radiochromatographic aspects of particular disciplines, eg
radiopharmaceutical quality control[2], have recently been reviewed the
coverage of these areas has been further restricted. The aim has been
to provide a limited insight for the general reader, rather than
detailed information for the specialist. To do otherwise would be to
restrict the space available for the description of other less well
reviewed areas.

2. THE TECHNIQUES OF RADIOCHROMATOGRAPHY

Any definition of the area of radiochromatography is of necessity as imprecise as that of chromatography itself. For the purposes of this article I have attempted to allocate the available space amongst the range of radiochromatographic techniques in approximate proportion to the extent of their use across a range of disciplines. Hence there are major sections on thin layer and high performance liquid chromatography, since these techniques serve many disparate areas of science. Conversely, the areas of gas liquid chromatography and electrophoresis fare less well, even though they are the mainstay of important specialist areas such as lipid and steroid research, protein chemistry and molecular biology . As yet there are no reported uses of supercritical fluid chromatography in the area , though this technique appears to possess good potential for radiochromatography. The next sections deal with the individual chromatographic techniques in some detail.

Planar chromatography.

Planar chromatography is a term coined to describe those techniques involving a significant two dimensional component. Thus thin layer, paper, preparative layer, and overpressure techniques are all included in this definition, which thus avoids unnecessary fragmentation of fundamentally related techniques. Of the planar chromatography techniques by far the most utilised is thin layer chromatography (tlc) and its derivatives[3,4,5,6] though paper chromatography is still used for some applications[7,8].

When applied to radiochromatography, planar techniques have many advantages with respect to those which are based on columns. Planar media, for example, are cheap and are used once only prior to disposal. Consequently they are well suited to radiochemical work, where persistent contamination of reusable media can present problems. In addition, in many radiochemical studies the primary requirement is for chromatographic separation, rather than for high resolution. It is in this area that planar chromatography, and thin layer chromatography in particular, possesses great advantages. The wide range of stationary phases available, combined with a virtually unrestricted choice of mobile phase compositions and pH values imparts excellent separation ability.

In our laboratories we regularly employ thin layer radiochromatography for initial evaluations of the radiochemical purity of labelled pharmaceutical agents and synthetic intermediates. By a judicious choice of stationary and mobile phases each labelled substance may be subjected to a wide range of different separation processes eg adsorbtion, partition, reversed phase, ion-exchange, ion-supression, hydrogen-bonding, ion-pairing, etc. Any one of these separation modes may reveal the presence of an unknown , and often unsuspected, impurity. Since thin layer eluents are rapidly prepared, used in small quantities, and are often stable upon storage, such an extensive analytical screen is both simple to employ, cheap and very rapid. In such a screen the thin layer radiochromatograms are prepared by concurrent development in all the selected chromatography systems and then subsequently quantitated by automated equipment. Tlc utilised in this fashion forms a very powerful, and advisable , pre-screen before detailed quantitation of impurity profiles by techniques possessing better resolution, eg gas liquid chromatography (glc) or high performance liquid chromatography (hplc). In this screening role tlc is unmatched for speed, convenience and reliability.

Another advantage of planar chromatography is that all the radioactivity applied to the chromatographic medium is available for detection. Even colloidal, particulate, or other insoluble components of a sample will be detected at some point on the radiochromatogram. Likewise, crystallisation, decomposition, or precipitation of a component induced by the chromatographic system, will be observed. This is a significant advantage of planar chromatography when compared to alternative column techniques such as glc or hplc. Indeed the ease with which the latter types of behaviour can be detected has played some part in supporting the view that planar chromatography in particular is prone to artifacts.

Since automated sample application equipment is commercially available, tlc is also well suited to applications involving the analysis of large batches of labelled samples. In these cases, many radiochromatograms may be prepared and then developed simultaneously. The radioactive components may then be detected in parallel by autoradiography or sequentially by an automated linear analyser.

A further unique advantage of planar systems, that of two dimensional chromatography, has been extensively employed in a range of applications where complex separations or the fingerprinting of complex mixtures are required.

One significant disadvantage of tlc is that volatile radioactive components, eg tritiated water, may not be detected by the technique.

Another disadvantage of tlc is low resolution. Thus, 20 cm thin layer plates achieve only around 500 usable theoretical plates compared with a typical 25 cm hplc column where the figure is around 20,000. This low resolution should be seen in the context of the resolution of commercially available radioactivity detectors most of which degrade resolution substantially when compared with conventional chromatographic detectors. Moreover, the development of high performance thin layer chromatography (hptlc) which delivers ca. 5000 theoretical plates/10 cm development, coupled with higher resolution linear analysers has gone some way towards closing this gap. Recently the development of overpressurised layer chromatography (oplc)[9,10] has more than doubled this resolution. Despite such developments it is nevertheless unlikely that planar techniques will deliver higher resolution than column techniques in the near future. Their strength remains their unique versatility in separation modes.

Recent applications illustrate the widespread use of radio-tlc. They include; the investigation of chlorophyll biosynthesis[11], assay of pentaerythritol and metabolites in blood[12], quantitation of nucleotides and nucleosides[13], assay of long-chain derivatives of coenzyme A[14], determination of amino acid decarboxylase activities[15], microanalysis of ^{123}I-labelled myocardial lipids[16], elucidation of the basic carbohydrate structures of gangliosides[17], separation of ^{125}I-labelled cyclic nucleotides by ion-exchange tlc[18], preparation of highly tritiated boar pheromone[19], and analysis of insect hormone metabolites[20].

Detection in Planar Chromatography.

Plate sectioning techniques. One of the earliest methods for the detection of radioactivity on planar chromatograms was to section the chromatogram, either by cutting or scraping off zones of interest. Subsequently the associated radioactivity was assayed by a suitable radioactivity counter, eg a liquid scintillation spectrometer. The technique is laborious and subject to many errors but is still in regular use. Indeed a number of automated systems for plate sectioning have been produced commercially.

Plate sectioning techniques have been used particularly effectively in radiopharmaceutical quality control applications such as the determination of radiochemical purity of a wide selection of ^{99m}Tc-labelled preparations[21]. Various adsorbents have been employed in this type of application, eg paper strips, glass-fibre sheets impregnated with silica or silicic acid etc. Detection in this application is simple, requiring only a well-type scintillation counter or, for higher activities, a simple radioisotope calibrator[22].

Recently, the advent of tlc radiochromatogram scanners, linear analysers[23] and radioactivity imaging systems, all of which are more convenient, automated, of good accuracy, and much more rapid, has severely reduced the popularity of plate sectioning methods.

Autoradiography. In its simplest form the autoradiographic
detection procedure consists of placing the dried chromatogram or gel
in close contact with a sheet of photographic film for a suitable
period of time. Under these conditions the radioactive emissions
from the chromatogram produce a latent image in the emulsion which
can be visualised by development and fixing of the film. The
procedure is extremely convenient, cheap, and, at least for weak
beta-emitting nuclides, has a resolution similar to that of the
chromatogram itself. The time required for autoradiographic detection
varies with the type and quantity of radiation and the type of film
used.

Due to the short range and low energy of tritium beta-particles
the detection of this isotope by autoradiography is difficult . For
example, tritium is detected with efficiencies several thousand times
lower than for ^{14}C by the typical X-ray films used in
autoradiography. The efficient detection of tritium therefore
requires specialised techniques. These include the use of
tritium-sensitive film which possesses a thin emulsion layer and does
not have a protective gelatine coating to prevent mechanical damage.
Because this latter layer is not present a far greater percentage of
the tritium beta-particles arrive at the emulsion; however the
absence of the layer does render the film more subject to chemical
and mechanical handling artifacts. An alternative detection technique
is to convert the tritium beta particle energy into light by the use
of a suitable scintillator applied to the radiochromatogram prior to
autoradiography. This technique, fluorography[24], is extensively
utilised in the biomedical and life sciences areas for the
autoradiography of electropherograms etc.

Scanning detectors. In their simplest form these detector
systems consist of a mechanical system for the transport of the
chromatogram past a fixed radioactivity detector or, conversely, for
the transport of the detector system over the surface of the
chromatogram. In either case, the utility of such systems is limited
by a compromise between sensitive detection of the radioactivity and
transit time. The detectors used are Geiger, scintillation crystal,
or gas-flow proportional counters. For years these systems were the
mainstay of radio-tlc analysis but they have now largely been
replaced by linear analysers. There is still a considerable use of
scanning systems of this type for the analysis of radiochromatograms
involving gamma emitters. In this case thallium doped sodium iodide
crystal detectors are often employed. Even in this application,
though, linear analysers can give better performance in many cases[25].

Spark chambers. Spark chambers exploit the ability of radiation
to cause a discharge between two wire grids held at a high potential
with respect to one another. The ionisation caused by the passage of

emitted beta particles close to or in the high potential region makes the probability of a spark in that region high. If, therefore, a planar chromatogram is placed close below such a grid, the positions of any labelled zones will be indicated by enhanced sparking. With good lightproofing of the system, these areas of high sparking can be observed visually or can be recorded by a camera set for time exposure. The spark chamber is applicable to many types of radiation but is particularly well suited to beta-emitters. Sensitivity is quite high and under optimal conditions chromatographic spots containing as little as 0.5 Bq of ^{14}C radioactivity can be detected with exposures of around 10 minutes. The linearity of spark chambers is not very high however, and consequently they are most often used in qualitative or semi-quantitative applications. For sensitive recording of the tritium isotope it is necessary to remove a protective anti-contamination film from the anode/cathode array in order to achieve high sensitivity. This opens the way for potential contamination of the detector.

Linear analysers. Over the last five years the assessment and quantitation of planar radiochromatograms has undergone a significant advance with the commercial introduction of a new detection system, the linear analyser. The technology underlying this type of instrument has been developed from the large scale computerised radiation detectors utilised in fundamental physics research. The analysers offer a rapid, simple and quantitative procedure for tlc, paper, gel electrophoresis, and related planar chromatographic techniques. A diagram of a typical linear analyser detector is given in Figure 1.

Central to the operation of these systems is a detector head consisting of a position sensitive wire detector[23] adjacent to, and parallel to which, is a coiled delay line. In use, the detector head, which is open at the bottom face, is placed longitudinally over the radiochromatogram and is continually purged with a slow flow of counting gas from a cylinder. When beta particles or other radiation from the radiochromatogram enter the detector head they produce ionisation in the counting gas. Because of the presence of a high potential on the detector wire this primary ionisation gives rise to a secondary ionisation avalanche in the vicinity of the detector wire. This event electromagnetically couples a signal into the nearby delay line at the position of the original radioactive emission. The induced signal in the delay line then propagates, with a fixed velocity, along the line in both directions until it reaches the ends and is detected. Since the detection systems used in linear analysers can detect time differences as small as one nanosecond[26], comparison of the difference in the time of arrival of the signal at opposite ends of the delay line can be used to calculate the position on the delay line at which the signal originated. Hence, the position on the radiochromatogram at which the initiating disintegration

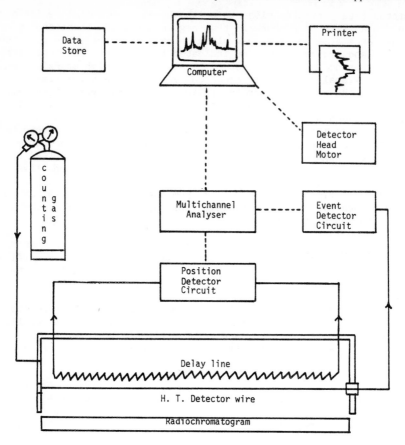

<u>Figure 1:</u> Diagram of one type of linear analyser. The detector head (not to scale) is shown sectioned vertically and parallel to the major axis of the radiochromatogram track

occurred is defined.

Since, unlike scanning systems, neither the detector nor the radiochromatogram is required to move in order to establish positional information, there is no compromise necessary between transit time and sensitivity. Linear analysers are therefore sensitive in comparison with the previous generation of scanners[27,28].

A wide range of radionuclides can be detected by linear analyser systems. Resolution is highest for weak beta-emitters or for electron capture nuclides such as ^{125}I for which the instrument detects the associated Auger electron emission.

Although open window detection is commonly used, the bottom of the detector head can be sealed against contamination by a thin film of mylar. For most beta-emitters, but not for tritium , this does not introduce a significant reduction in sensitivity. Even for tritium a thin gold-deposited foil is available.

Radioactivity imaging systems. Four different types of two-dimensional radiochromatography imaging systems are under development. The first involves the use of a linear analyser such that an image of the radioactivity detected on a two- dimensional radiochromatogram is achieved by computerised superposition of a series of one-dimensional readings. These are acquired sequentially by stepping the detector in small increments across the second dimension of the radiochromatogram. Similar systems which utilise delay line technology to produce two-dimensional images are under current development. In one of these systems the detector consists of a square crossed matrix of detector wires which are sampled by delay lines at each edge. The device uses delay lines fabricated by thick film technology and initial evaluations of sensitivity and resolution look promising.

The second type of imaging system utilises the spark chamber principle, but detects the sparking by the use of an image intensifying system linked to a video image analyser. Alternatively the image may be detected by a charge coupled device (ccd) camera which outputs to an image analysis computer. This system, which uses a soft argon-neon sparking gas can detect as little as 0.2 Bq of ^{14}C radioactivity in a 40 minute period and is commercially available. Since a mylar window is used the system is not applicable to tritium.

In the third type of imaging system, the radiochromatogram is placed in contact with a thin sheet of scintillator plastic and the scintillations produced are detected as previously by an image intensification system or by a Peltier-effect cooled ccd camera.

Currently this system is not sensitive enough for the detection of tritium. However, a variant on the above technique has been described which is suitable for the detection of both tritium and carbon-14 on tlc plates. The system utilises an imaging apparatus consisting of a two stage microchannel-plate image intensifier, a low-lag vidicon and an image frame processor. This detector, operating at room temperature, can image and quantitate the scintillations arising on tlc plates impregnated with the scintillant 2,5-diphenyloxazole. Under these conditions detection limits of 0.2 and 8 Bq/mm² are quoted for carbon-14 and tritium respectively[29].

Finally, a device consisting of a sequence of multi-wire proportional chambers and position sensitive detectors has been described[30] for the video-imaging of tlc plates and slab electropherograms bearing ^{3}H, ^{14}C and ^{32}P. Radiochromatogram zones comprising 25, 3.3 and 2.5 Bq respectively of these isotopes could be imaged and discrimination of all three isotopes was possible. Modification of the above system by the use of a multi-step avalanche chamber detector increases the spacial resolution for the higher energy nuclides, though at the cost of some sensitivity[31].

Artifacts in Planar Chromatography.

In view of the wide variation which is possible in the stationary phase, mobile phase, sample application solvent and elution conditions it is not surprising that a wide range of artifacts can be observed in planar chromatography. Many of these artifacts are not peculiar to radiochromatography. Rather, the ease of detection of labelled compounds renders the observation of chromatographic artifacts particularly likely. Typical tlc artifacts have been discussed previously[1,32].

A few artifacts are specific to the area of radiochromatography, particularly those associated with radiochemical detection systems or with the application of exceptionally low sample masses to the chromatogram. This latter is often the case in the analysis of high specific activity compounds.

In our experience, linear analysers are particularly prone to display extremely narrow apparent peaks of radioactivity if the chromatograms have been eluted with solvents containing acidic compounds of high dielectric constant, eg formic acid or diluted mineral acids. Such artifact peaks can persist for many hours on apparently dry chromatograms. Similar peaks can arise from markings added to the chromatogram in pencil, from other conductive sharp edges or from dust particles. This type of artifact can sometimes be eliminated by the use of a reduced detector voltage.

Autoradiography is also prone to artifacts arising from phosphorescence, chemiluminescence, or from chemical or residual eluent attack on the emulsion. This latter behaviour is particularly evident with tritium-sensitive films which do not possess an anti-scratch gelatine coating to cover the sensitive emulsion layer. Artifacts arising from such chemical interaction can be easily detected by comparison with a similar chromatogram generated from the unlabelled analyte.

Liquid Chromatography.

Open column liquid radiochromatography continues to be used in preparative applications, though increasingly its function is being replaced by two alternative techniques, high performance liquid chromatography, which is discussed in the next section, and flash chromatography[33]. This latter is an extremely useful large scale preparative technique involving large bore glass columns charged with a fine adsorbent through which the mobile phase is forced under moderate gas pressure. Both ordinary and reversed phase adsorbents are commercially available and this flexibility combined with the rapid separation time makes the technique extremely useful in radiosyntheses from the one to twenty millimole scale[34,35].

High Performance Liquid Chromatography

The popularity of high performance liquid chromatography[36,37] with respect to other chromatographic techniques has increased enormously over the last decade and this change has resulted in a corresponding expansion in the use of hplc in a wide variety of radiochromatographic studies. Undoubtedly the popularity of the technique is related to its versatility. Thus an extremely wide range of compound classes may be analysed by the technique rapidly, with high resolution, without derivatisation, at room temperature and without significant constraints on molecular weight, polarity or ionisation state. These advantages are of particular importance in the life sciences area where the analysis of complex mixtures often including unstable compounds is common. Recent applications of the technique include; studies of enzyme function, specificity, or activity[38,39,40], sterol and insect hormone biosynthesis[41,42] studies with arachidonic acid metabolites[43,44,45], the analysis of labelled PTH amino acids[46], identification and determination of tritiated metabolites of enkephalins[47], analysis of sugars[48], and the isolation of mono-and polysaccharides from labelled biomass[49].

An increasingly important area is the analysis and purification of labelled proteins , for example; the isolation of ^{125}I-labelled polypeptide and protein hormones[50,51], isolation and analysis of labelled antibodies[52], and studies of ^{35}S-labelled glycoproteins[53].

Whilst the range of applications of radio-hplc in these areas is indeed very large, other areas are also utilising the technique to an ever greater extent. Hplc now plays a primary role in the purification and quality control of radiopharmaceuticals[54], labelled fine chemicals[55], labelled pharmaceuticals[56,57,58] environmental agents[59] etc, whilst the preparation of compounds labelled with short-lived positron emitters[60,61,62,63,64] or of other unstable labelled compounds[65] is almost completely dependent on the technique. For particularly short-lived isotopes even hplc-based syntheses are too slow and resort must be made to micro-column techniques[66].

Radio-Hplc Equipment.

The configuration of a typical radio-hplc is shown in Figure 2. It comprises four fundamental parts, of which only the detector systems are peculiar to radiochemical work. The various components are described below.

Hplc pumping systems. The type of pumps employed in radio-hplc varies according to the application. In most instances the pumps comprise a dual or ternary system which confers on the radio-hplc the ability to rapidly select mobile phase compositions and to perform gradient elution. This latter technique is particularly important in radiochemical applications where complete elution of the radioactivity from the chromatographic system must be assured. For analytical applications or for very small scale preparative work with high specific activity compounds, pumps with a range of flow rates of around 0.1 to 10 ml/min are ideal since this range is suited to chromatographic columns of around 3 to 8 mm internal diameter. For preparative applications however, for example the purification of a few hundred milligrams to a few grams of ^{14}C-labelled compound, pumps with a maximum flow rate of around 20-60 ml/min are desirable. This enables the use of preparative columns with internal diameters ranging from 10-50 mm. A number of commercial systems are available which allow a rapid switching between analytical and preparative modes. Hplc pumping systems with very small flow rates suitable for microbore or capillary work have little application as yet since the utilisation of these techniques in radio-hplc is hampered by the resolution of radio-detector systems.

Sample injector systems. For the majority of applications in radio-hplc the sample is manually introduced onto the chromatographic column by means of a syringe and loop injector. These systems consist of a multi-port valve including a fixed volume loop. The loop can be

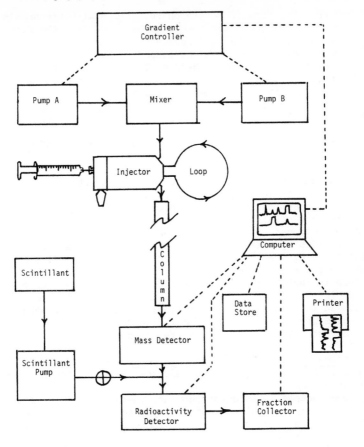

<u>Figure 2:</u> A typical radio-hplc configuration. Fluid connections are shown as solid lines whilst control or data connections are shown as broken lines

isolated from the flow of mobile phase, filled with the sample solution to be injected, and then switched back into the flow. The sample in the loop is then carried onto the chromatographic column by the flow of mobile phase from the pumps. Probably the most common injector of this type is the Rheodyne 7125, though others are manufactured by Valco, Waters etc. Loop sizes from a few microlitres to several millilitres are commercially available. These suffice for the majority of applications. However for large scale preparative work particularly with compounds of low solubility, either a large loop must be constructed from wide diameter stainless steel tubing or the sample must be loaded by introduction through one of the pump inlets.

An extensive range of automated sample injector systems is commercially available and these are extremely useful for the repetitive injection of single samples or, more commonly, for the automatic sequential injection of large numbers of different samples. The mode of operation of these autosampling systems varies with the make, but most systems will hold from 40 to 100 separate sample vials whilst allowing the injection of volumes from around five microlitres to around two millilitres. Recent autosamplers are extremely flexible in their operation and will inject from the rack or tray in any specified order. In addition elements of robotics are being introduced into autosampler operation; recent models provide for a range of manipulations of the sample eg dilution, derivatisation or even solvent extraction, prior to injection.

Chromatographic columns and stationary phases. A very wide selection of chromatographic columns are available for hplc work, the majority are simple packed columns of stainless steel, but the use of cartridge systems is becoming popular. These latter have some merit in an area such as radiochromatography where cross-contamination could occur and where disposal of columns used for high level work is either problematic or time consuming. Some mention should be made of commercial pre-packed medium pressure glass columns which are semi-disposable and which often prove extremely useful by providing high sample loading along with medium resolution in preparative work. Most high pressure, high resolution preparative columns are constructed of stainless steel with internal diameters of 10 to 50 mm and with lengths of 100 to 500 mm. The cost of these columns packed with small diameter (5 micrometre) stationary phases is high but, in our experience, the resolution is hardly inferior to a similar analytical column and column lifetimes are generally quite long.

Although microbore hplc techniques have been available for several years, they have found little use thus far in radiochromatography. Their extremely high resolving power has been utilised, however, in the separation of deuteriated compounds from their protiated analogues[67,68].

The type of stationary phase utilised in radio-hplc varies enormously,though small particle silica or C18 reversed phase silica systems are by far the most popular in a very wide range of applications. A few studies have been reported which utilised chiral stationary phases for the resolution of enantiomers of chiral labelled compounds[69]. Interestingly, the separation of ^{14}C-labelled enantiomers of nicotine in the presence of the unlabelled materials, using a totally achiral column and mobile phase, has been reported[70].

·Mass detectors. In addition to some form of radioactivity detector, most radio-hplc systems employ at least one other detector such that radioactive components can be identified by properties other than their radioactivity. These detectors are often referred to as mass detectors since they usually provide an estimate of the quantity of individual components by exploiting a property, eg ultraviolet absorbance, which is proportional to the mass of component present[37]. By far the most common mass detectors are ultraviolet absorbance detectors,though other types exploiting electrochemical oxidation and reduction, conductivity, refractive index, fluorescence etc, may be used for particular types of analyte. The ultraviolet absorbance detectors employed in radiochemical applications are very varied, though for most non-preparative applications simple detectors with flow cells possessing path lengths of 5-20 mm and with sensitivity ranges from around 0.01 to 2.0 absorbance units for full scale deflection are quite suitable. The wavelength range of such detectors is usually from around 200 nm upwards and this also will suit most applications. For preparative applications on the millimolar scale, flow cells of 0.1 to 0.5 mm path length are desirable to avoid detector saturation.

Modern, more sophisticated, detectors allow the recording of an ultraviolet spectrum for selected components during the chromatography, and this can be useful for identity confirmation. In our laboratories we routinely employ ultraviolet photodiode array detection in conjunction with radioactivity detection. This combination of detectors enables the maximum amount of information to be derived from each chromatographic analysis. Since all the raw data accumulated during the chromatography is stored in the associated computer system, a wide range of post analysis data processing techniques can be employed to ascertain peak purity and identity[71], to select the most suitable wavelength for future quantitation, or merely to select a suitable display format for reporting. Before leaving the subject it should be noted that ultraviolet absorbance or other mass detectors are usually placed prior to any radioactivity detector employed in the radio-hplc. In this way any band spreading introduced will not degrade the performance of the mass detector. Obviously this practice must be followed if homogeneous flow detection techniques are employed.

Hplc Radioactivity Detection.

Fraction collection/counting . The simplest form of
radioactivity detection for hplc is to collect fractions and to assay
the associated radioactivity using a suitable off-line detector eg a
liquid scintillation spectrometer. This technique is accurate,
simple, direct and requires no specialised apparatus other than that
routinely available in most radiochemical laboratories. Indeed it is
probably still the most commonly used technique in laboratories in
which the use of radio-hplc techniques is limited. When the use of
radio-hplc is extensive however, the procedure becomes expensive both
in terms of liquid scintillation supplies and in counter utilisation
time[72]. Moreover, since a typical study using such techniques, say
in the life sciences area, generates around 50 to 200 fractions per
chromatographic analysis, the radioactivity data does not become
available to the operator for several hours even if short counting
times can be employed and access to the liquid scintillation counter
is unrestricted.

Quantitative autoradiography . Whilst autoradiographic
techniques are ordinarily associated with planar chromatography, it
has nevertheless proved possible to apply them in the radio-hplc
field[73,74]). Thus a micro-fraction collector can be modified to
deposit aliquots of chromatographic effluent into depressions formed
in a sheet of non-wetting fluorocarbon film. Subsequent evaporation
of the solvent, transfer of the residual droplets to a paper sheet
and autoradiography then allows quantitation by densitometry. By the
use of this technique aliquots containing 17 Bq of radioactivity can
be assayed in six hours.

Flow detection methods

Many of the limitations of off-line detection are removed by the
use of flow detection. Ideally, a radioactivity flow detector should
be sensitive, should yield immediate access to data, should not
compromise chromatographic resolution, and should be easy to operate.
In reality however, all flow detectors are subject to constraints
imposed by the random nature of radioactive decay process, by a range
of instrumental limitations and by the quality of the chromatographic
performance. Indeed eight separate parameters controlling the limit
of determination of a typical flow detector have been described[75].
Adequate design and operation of such a flow cell is therefore a
question of skilful optimisation especially since several of these
factors are interrelated.

Perhaps the most important factors for the successful design of
a flow detector are the flow cell volume and the cell transit time.
As far as sensitivity is concerned these two parameters should both

take large values. Thus, maximum sensitivity will be obtained with a large flow cell to incorporate the maximum quantity of radioactivity within the detector, whilst a long cell transit time ensures that the radioactivity in the cell is summated over an extended period. In practice large cell volumes are precluded by the need to maintain good chromatographic resolution and the beneficial effect of long transit times are significant only if the detector has a very low background count rate. Consequently, both parameters are adjusted to provide a maximum signal to noise ratio. An alternative strategy, based upon post-column segmentation and on-line extraction of the eluate with a suitable immiscible liquid scintillator, has been described[76]. This system allows the post-run re-introduction of chromatographic regions of interest into the flow cell at a lower flow rate more commensurate with sensitive detection.

The above discussion of sensitivity applies to all flow detectors employing a single detection cell. If a linear series of low volume detector cells are utilised, however, the detection sensitivity can be increased by computerised summation of the counts obtained as a radioactive component passes sequentially through each of the cells in the array. Such synchronised, accumulating radioactivity detectors have been described for use in radio-glc, hplc and tlc applications[77,78].

Quite different designs are required for the detection of low energy beta and high energy gamma emitters. Diagrams of typical beta and gamma detectors are shown in Figure 3.

<u>Beta particle detectors</u>. Typical beta-detectors employ a radiochemical flow cell which consists of a coiled fluorocarbon tube positioned between two sensitive photomultiplier systems. The output from the photomultipliers is passed through a fast coincidence filtering system, which reduces thermal noise, to a ratemeter and from thence to either to a data system or to an analogue recorder.

The detailed designs of flow cells vary according to the manufacturer but usually the coiled fluorocarbon tube is embedded in a scintillator plastic to increase the efficiency of detection for high energy beta-emitters and is shielded by a metal block to reduce background count rates. Cell volumes vary from around 10 mm^3 to 6 cm^3.

Two modes of detection can be employed for the detection of beta emitters with low to moderate energies. In the first mode, heterogeneous detection, the flow cell is packed with a solid scintillant which emits photons, for detection by the photomultipliers, when in contact with beta radioactivity. In the second detection mode, homogeneous detection, the coil is left empty

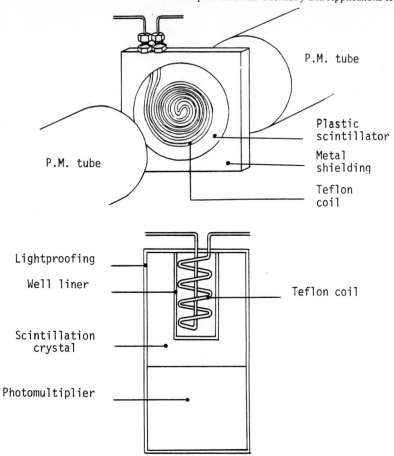

Figure 3: Typical flow detectors for radio-
hplc. Top cell for homogeneous/heterogeneous
beta counting. Bottom cell for detection of
γ emitters

and a liquid scintillant solution is introduced into the effluent
flow prior to the detection coil via an ancillary pump and a mixing
tee.

   Heterogeneous counting . A wide variety of scintillator bead
sizes and compositions have been proposed for use in solid
scintillator systems. Thus, carbon-14 can be detected with
efficiencies as high as 70% by the use of cerium impregnated lithium
silicate glass scintillator beads, the size distribution of which has
been selected to approximate the mean carbon-14 beta particle range
in the mobile phase[79]. Even higher efficiencies of 85% have been
claimed for fine yttrium-silicate glass beads. Sulphur-35,
phosphorus-32 and other medium to high energy beta emitters can be
detected efficiently by most commercial glass based scintillation
beads.

   A serious limitation of all the solid scintillators is the low
detection efficiencies for tritium; around 1% for glass based
scintillators, 5% for europium activated calcium fluoride[80], and at
best 8% for fine yttrium-silicate beads. This low efficiency is
associated with the particularly short range of tritium beta
particles most of which are absorbed by the mobile phase and do not
reach the scintillation bead surface where they could instigate the
scintillation process. For the detection of this isotope therefore,
a large amount of radioactivity must be injected and this often leads
to irreversible adsorbtion of a very small but significant fraction
of the labelled compound on the scintillator beads. Once retained in
close proximity to the scintillator in this fashion the detection
efficiency for beta particles from the adsorbed material increases
radically, giving rise to an increased background count in the cell.
We have encountered this memory effect for the tritium isotope many
times in our laboratories and can suggest no solution other than to
replace the scintillator beads.

   One very useful feature of heterogeneous detection is that none
of the injected radioactivity is wasted. It is therefore particularly
useful in multi-step synthetic and preparative applications where
sensitivity is of little concern but where the commitment of say
5-10% of the products to homogeneous detection via a stream splitter,
at each stage, would be unacceptable. Moreover, since no liquid
scintillant needs to be mixed with the effluent,compatibility of the
scintillant and mobile phase is not a problem . High concentrations
of organic solvents, complex and concentrated buffers and other
additives, can consequently be employed in the mobile phase.

   Homogeneous counting. In this technique the detector cell coil
is left empty and a liquid scintillant is used instead of the solid
scintillation beads. This scintillant is introduced into the mobile

phase by a metering pump prior to the effluent entering the flow
cell. Since the scintillation system is now homogeneous, the
efficiencies of detection for low energy beta emitters such as
tritium are high, generally of the same order as for a liquid
scintillation counter, eg ^{3}H ca. 50%, ^{14}C ca. 80%. Even electron
capture nuclides such as ^{125}I may be detected with reasonable
efficiency of around 50%.

With a well designed homogeneous flow cell detection limits are
excellent. Thus, peaks containing as little as 1.0 and 0.5 Bq of ^{3}H
and ^{14}C respectively can be detected with good confidence. In
addition to this high sensitivity, homogeneous detection has many of
the other advantages of liquid scintillation spectrometry such as the
ability to differentiate isotopes by their scintillation spectra.
Since most modern flow detectors incorporate resident microcomputers,
it is therefore possible to perform multiple isotope experiments and
to obtain separate radiochromatograms for each of the isotopes from a
single injection. Furthermore, since the variation in quenching
which occurs during gradient operation can be determined from a blank
gradient chromatogram, modern computerised flow detectors can be used
for accurate quantitative studies.

Although homogeneous techniques have many advantages they are
nevertheless subject to a number of drawbacks. One significant
limitation of homogeneous detection systems is the requirement to
ensure compatibility of the chromatographic mobile phase with the
scintillant, otherwise gelling or outright precipitation of mobile
phase buffer components can occur. Moreover, unless a stream
splitting system is utilised, homogeneous detection systems are
unsuitable for preparative work or for other hplc separations
requiring isolation of radioactive components.

A special type of flow detection is possible for high energy
beta-emitters. With these nuclides the Cerenkov photon emission,
which occurs when the emitted beta particles travel at speeds faster
than the speed of light in the mobile phase, can be detected directly
without the need for any scintillator. Efficiencies of ca. 30% are
possible with extremely low background count rates.

High-energy flow detectors. Although the foregoing beta
scintillation detectors can also be used, with varying efficiencies,
to detect gamma-emitters, these nuclides are better detected by the
use of an external scintillation crystal rather than an in-stream
scintillant. This is so not only because better counting efficiency
can be achieved for high energy gamma-photons but also because memory
effects can be eliminated. For example, common phenomena such as
iodine adsorption or the in situ formation and trapping of technetium
colloids are both avoided.

Many gamma-flow detectors are still purpose built and generally consist of a cell constructed from coiled fluorocarbon tubing set within a thallium-doped sodium iodide well-detector. For energies less than some 50 keV however, a cadmium-tellurium semiconductor-crystal detector is sometimes used. The output from the well-detector photomultiplier is then amplified and directed to a suitable quantitation or analysis device, eg a digital ratemeter, frequency-voltage converter or a multichannel analyser. Many applications use the latter instrument, operating in the multiscaler mode to acquire counts per unit time. Since most multichannel analysers have associated computers, the output from the detector can be arranged in a wide range of formats.

Recently, small and inexpensive high energy flow-detectors have become available. These are largely aimed at the increasing market arising from the widespread use of the iodine isotopes and phosphorus-32 in the life sciences, particularly in molecular biology and immunology applications.

Radio-Gas Liquid Chromatography

Gas liquid chromatography was amongst the earliest instrumental high resolution chromatographic techniques to be introduced, the first application taking place in the years 1950-1955. Over the subsequent period many instrumental refinements and detector innovations have been introduced. These have ensured that glc techniques continue to be amongst the most powerful and popular in the field of analytical chemistry[81,82]. The first application of glc techniques to labelled compounds[83] occurred in 1955, and over the subsequent thirty years or so, many hundreds of other applications have been described. The technique has been reviewed many times; particularly comprehensive overviews are referenced[84,85]. Whilst, currently, the use of radio-hplc is undergoing a large expansion, radio-glc utilisation has not increased much over the last decade. Nevertheless, in a number of research areas radio-glc still continues to be the method of choice. Thus, for the analysis of thermally stable volatile compounds, or for those compounds which can be conveniently converted to volatile derivatives, radio-glc continues to have many applications.

Recent examples of such applications include; determination of androgen reductase activity[86], studies of the curing of epoxy-resins[87], determination of mono-ethanolamine in crops[88], studies of cholesterol biosynthesis rates[89], and the analysis of tritiated oils[90].

In our laboratories we have found glc to be particularly applicable to the initial phases in the preparation of labelled compounds, since these early synthetic steps usually start from low molecular weight, often volatile, precursors.

Equipment for radio-glc. For the majority of radio-glc applications simple packed-column, temperature-programmed chromatographs have been used. A typical radio-glc system is shown in Figure 4. Mass-detection has largely been by flame ionisation detector, though powerful hybrid techniques involving the use of an ancillary mass spectrometer have been described[91,92,93].

Although capillary and narrow bore columns have been increasingly used in routine glc analysis, their applications in radio-glc are still limited. This is largely due to the dearth of commercial gas-flow radioactivity detectors possessing sufficient resolution to match the performance of current capillary columns. Recently, a commercial detector which utilises a micro-scale platinum catalysed oxidation/reduction tube and which has resolution compatible with capillary glc techniques has become available.

Radioactivity detection. The simplest method of radioactivity detection is to count aliquots of the column effluent using a liquid scintillation counter. Collection of the effluent has been achieved by a range of methods amongst the simplest of which are collection on absorbent cotton wool[90], on coated p-terphenyl crystals[94], or directly into a scintillation vial after oxidation by a flame ionisation detector[95]. In the latter method both tritium and carbon-14 can be detected with good efficiency provided that a heated collector tube is used.

In addition to manual collection, semi-automated effluent scrubbing systems which collect samples for subsequent liquid scintillation counting have been described[96].

Flow detectors. Two types of radioactivity flow detectors for radio-glc are in common use. They are based upon gas-flow proportional counters or upon ionisation chambers. Both these types of detectors are sensitive to temperature changes, and therefore not compatible with temperature programming. It is usual, therefore, to arrange for the conversion of radioactive components eluting from the column into non-condensable gases. Flow detection is then carried out at room temperature. Most flow detectors use a heated copper oxide oxidation tube to convert ^{14}C- and ^{3}H-labelled compounds to labelled carbon dioxide and tritiated water respectively. If ^{14}C alone is to be determined then the water is removed with a tube of desiccant prior to the gases entering the detector. If on the other hand tritium is to be detected, the tritiated water is converted to hydrogen by an iron-chip furnace. A clamp and flow diverter valve has

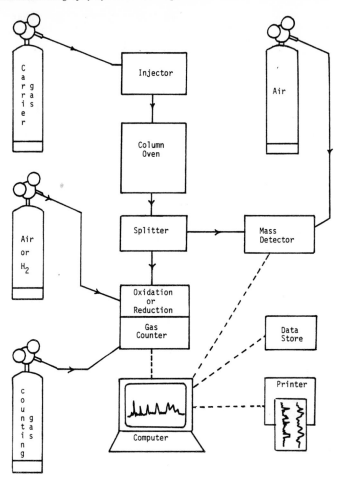

Figure 4: A typical radio-glc configuration. Gas connections are shown as solid lines whilst control or data connections are shown as broken lines

been described which reduces the need for frequent replacement of the
copper oxide when many samples, or when samples containing large
quantities of oxidisable materials, are analysed[97]. Recent
developments in small-scale platinum furnace technology have obviated
this difficulty, at least for the ^{14}C isotope.

Undoubtedly the most popular detectors are gas-flow proportional
counters. Variants of such detectors have been constructed which are
suitable for fused silica capillary operation[98], or for integration
into a mass spectrometer data system[92]. A gas-flow proportional
detector has been constructed, from commercially available
components, which enables the quantitation of tritiated steroids by
capillary radio-glc. Quantities as small as 13 Bq of ^{3}H radioactivity
could be detected well by this system, whilst the resolution of the
capillary column was only slightly degraded[99].

A synchronised accumulating detector has been described[100] which
employs five sequential small-volume gas-flow proportional counters
to increase detection sensitivity and reduce background.

Although ionisation chambers were utilised in the early stages
of radio-glc development, their use has subsequently decreased. One
particularly useful exception, however, is the modification of a
commercial flame ionisation detector to operate as an ionisation
chamber[101]. The resulting detector is not very sensitive; the limit
of determination for tritiated aromatics is around 40 kBq.
Nevertheless, the simplicity of the modification makes the use of
such a detector for preparative and radiosynthetic use very
attractive.

Radio-Electrophoresis

Gel electrophoresis techniques are fundamental to much of
current research in the biosciences and, though not strictly
chromatographic, they deserve some separate mention. In the basic
electrophoresis technique molecular species, usually macromolecules,
are separated according to their molecular charge. The separation
takes place during migration through a gel matrix, cast in the form
of a rectangular slab or cylinder, across opposite ends of which a
large voltage gradient is applied. Gels for charge separation are
usually cast from agarose, since this polysaccharide forms gels with
large pore sizes. If, however, gels are prepared by co-
polymerisation of acrylamide and bis-acrylamide, the average pore
size of the gel can be adjusted such that molecular sieving
accompanies the separation by charge. By a suitable choice of pore
sizes, co-electrolytes etc, such systems can be used for studies of
protein molecular weights, protein aggregation and so forth.
Procedures for the preparation of tritium-labelled standard

calibration proteins for such studies have been described[102].

A wide range of modified gel electrophoresis techniques are in use. Thus, the production of polyacrylamide gels across which the pH has been made to vary linearly, produces separations in which the various proteins are separated according to their isoelectric points. This is the isoelectric focussing technique; it gives extremely good resolution of many protein mixtures. Agarose gels may also be modified by the inclusion of antisera against specific proteins or protein classes. This modification gives rise to a plethora of related immunoelectrophoresis techniques which are used to identify and quantitate the respective protein antigens. The two dimensional gel electrophoresis of protein mixtures, including some aspects of radio-detection, has recently been reviewed[103].

In addition to its use in the separation of proteins, gel electrophoresis is a fundamental technique in the analysis of nucleic acid fragments. Here the driving force for the electrophoretic separation is the uniformly distributed negative charges associated with the sequence of ionised phosphate groups along the molecular backbone of the fragment. Using a range of gel porosities nucleic acid fragments from as small as 20 base-pairs to fragments around one thousand times larger (20kb) can be separated. Beyond this size the maximum pore-size of agarose gels becomes limiting, though the use of pulsed field gel techniques[104] can extend the upper limit of analysis to over one million base pairs.

Small fragments of up to 1kb can be separated by polyacrylamide gel electrophoresis. This technique is extensively used in DNA sequencing for the separation of DNA sequence-fragments which have been coupled with $^{32}P-$ or $^{35}S-$ labelled A,G,C or T nucleotides.

Radioactivity detection. Although beta radioactivity scanners can be usefully applied to the quantitation of two- dimensional gels[105], by far the most common detection technique for radio-electrophoresis is autoradiography. In a typical procedure the gel electropherogram is dried prior to radioactivity detection[106]. Commercial equipment which uses heat and reduced pressure to dry the gel onto a rigid medium such as a paper sheet is available for this operation. Subsequent exposure in contact with a suitable autoradiographic film yields the autoradiogram. For the detection of ^{14}C, ^{32}P, ^{35}S sheet X-ray film is used. For the detection of tritium it is usual to employ fluorographic techniques[107]. These procedures consist of impregnating the electropherogram with a solution of an appropriate scintillator, eg 2,5-diphenyloxazole, prior to detection of the scintillations using X-ray film. A range of competing fluorographic enhancers are marketed by the major radiochemical suppliers. The sensitivity of fluorography can be increased by exposing the film to the electropherogram at low temperatures (-70 degrees). By a two step process of autoradiography and then fluorography distinction can be made between ^{3}H-labelled zones and those labelled with higher energy nuclides[108].

Several flow detectors have been constructed for the detection of ^{32}P, ^{14}C, and ^{3}H-labelled compounds during capillary isotachophoresis[109]. No commercial detectors are available for this application however.

3. ISOTOPE EFFECTS IN RADIOCHROMATOGRAPHY

The fact that isotopic substitution can affect chromatographic retention has been known for many years and has formed the basis of a number of enrichment processes for metal isotopes[110]. Although much of this initial work involved metals and utilised ion-exchange techniques, it was soon apparent that the phenomenon also applied to isotopically substituted organic compounds and was observed in a wide range of chromatographic techniques. Indeed, so many instances of the fractionation of large organic molecules containing isotopic labels had been reported by 1966 that an extensive review was called for[111]. Since this date chromatographic developments have resulted in greatly increased resolving power for most techniques. Isotopic fractionation is therefore easily, and often unexpectedly, observed without specialised apparatus.

Since glc yielded the first high resolution chromatographic separations it was to be expected that the earliest pronounced isotope effects would be observed by this technique. As early as 1957, the glc fractionation of cyclohexane and perdeuteriated cyclohexane was reported[112], and in the same publication the authors noted a degree of additivity in the resolution of various multiply-tritiated isotopomers from the unlabelled compound. More recent studies with tritiated compounds have confirmed these observations[113]. This approximate additivity of isotope effect with increasing isotopic substitution is also commonly observed in tlc and hplc studies. There are, however, significant deviations from the rule.

By 1975 the fractionation of many small molecular weight isotopomeric mixtures had been achieved and the data reviewed[114]. Contrary to predictions based upon mass considerations alone, in the great majority of separations the heavier isotopomer elutes from the column first. This implies a more extensive interaction of the unlabelled isotopomer with the stationary phase and has been ascribed to a range of different but related processes occurring in the condensed phase. A general theory of such isotopic effects has been elaborated[115] and reviewed[116].

Following the development of gc-ms isotopic dilution analysis techniques, the observation of isotopic fractionation has become a common event. Occasionally, especially with capillary techniques, the fractionation is so complete that the ms-detector can be dispensed with, and the isotopically substituted compound utilised as an ordinary internal standard. For example, the complete separation of tetradeuterio-1,2-dibromoethane from the unlabelled material has been utilised in the glc analysis of the unlabelled pesticide in crops[117]. Interestingly, the simplifications introduced into the calculation by the presence of the heavy bromine atoms in this molecule has made comparison of the theoretical and observed glc isotope effect particularly easy in this case[118].

Chromatographic isotope effects are also observed in tlc separations. The effects are particularly large when isotopic substitution is in positions close to or adjacent to a basic nitrogen atom. This is the case when labelled compounds are prepared by methylation of amino functions with trideuterio- or tritritio-methyl iodide; the latter reagent is commonly used in the production of high specific activity radioligands for receptor binding studies. These, and similar, studies have produced many examples of large isotope effects for methadone[119], imipramine[120], oxymorphone, morphine, dihydromorphine, lysergic acid and mianserin[121] and bepridil[122]. In all these cases the chromatographic stationary phase used was silica and the unlabelled compound was observed to be less retained than its deuterated or tritiated isotopomers. The increased retention of the labelled compound has been ascribed[119] to the base strengthening effect of isotopic substitution on the nearby nitrogen atom; this stronger basicity being reflected as increased interaction of the nitrogen function with the acidic sites on the silica stationary phase.

Since hplc techniques are now used extensively for work with isotopes there are an increasing number of reports of isotopic fractionation. Often, with compounds containing several isotopic atoms per molecule, the separations are particularly large. Thus, argentation hplc has been used for the separation of polydeuteriated and polytritiated isotopomers of arachidonic acid metabolites from their unlabelled precursors[123]. The separations reported[124] yielded isotopically enriched prostaglandins, leukotrienes etc, which were suitable for use as gc-ms internal standards or radioimmunoassay radioligands. Although the above preparative fractionation was achieved on silver-loaded columns there are many examples of similar separations on simple reversed phase stationary phases. Carotenoids and perdeuteriated carotenoids for example, may be completely resolved on simple 25 cm C-18 columns[125] whilst under similar conditions chlorophylls of varying degrees of deuteriation could also be separated[126]. For the great majority of the isotopic fractionations observed using reversed phase hplc, the deuteriated or tritiated isotopomer is less retained than the unlabelled compound. Moreover, as with glc, the isotopic effects are additive; increasing numbers of isotopic atoms leading to increasingly early elution.

The mechanism of isotopic fractionation on reversed phase hplc has been discussed and ascribed to either differential perturbation of molecular vibration modes by the stationary and mobile phases, or to Van der Waals interactions of differing strengths between the stationary phase and the various isotopically labelled species[127]. This latter idea has been further developed by Baweja[125,126], who ascribes the separations to differential Van der Waals interactions between the C-H or C-D bonds of the analyte and the surrounding molecules on the stationary phase. Since the vibrational frequency of a typical C-H bond (ca. 3300-3000 cm-1) is larger than that of a typical C-D bond (ca. 2200-2000 cm-1) the electromagnetic field strengths induced in molecules surrounding the vibrating bond are larger for the hydrogen isotopomer than for the corresponding deuteriated molecule. As the induced fields lead to an increased interaction between the analyte and the stationary phase, the unlabelled compound is retained more strongly than the deuteriated analogue. Presumably an extension of the same argument would be advanced to explain the fractionation of unlabelled and tritiated isotopomers.

In certain cases the order of elution differs from that expected from the above discussion. Thus, chlorpromazine bearing a tritiated methyl group was more retained on reversed phase hplc than was the unlabelled compound[128], whilst deuteriated methadone eluted later than methadone[119] In both these cases, however, other factors affected the reversed phase interaction. In the former case the separation was highly pH-dependent, whilst in the latter case the authors cite n.m.r. data in support of a conformational change occurring in one of the isotopomers[119].

4. REFERENCES

1. J. Chromatog. Library, Vol 41 : 'Radiochromatography', Roberts T R, Elsevier, 1978.
2. 'Analytical and Chromatographic Techniques in Radiopharmaceutical Chemistry', eds Wieland D M, Tobes M C Mangner T J, Springer-Verlag, 1986.
3. 'Techniques and Applications of Thin Layer Chromatography', eds Touchstone J C, Sherma J, John Wiley and Sons Ltd, 1985.
4. 'Thin Layer Chromatography', Hamilton R, Hamilton S, John Wiley and Sons Ltd, 1987.
5. 'Practice of Thin Layer Chromatography', 2nd Ed., Touchston J C, Dobbins M F, John Wiley and Sons Inc., 1983.
6. 'The Radiochromatography of Labelled Compounds': Review 14, Sheppard G, The Radiochemical Centre, Amersham, 1972.
7. Filer C R, Ahern D G, J. Lab. Comp. Radiopharm., 24, 615- 622, 1987.
8. Dangelmaier C A, Daniel J L, Smith B J, Anal. Biochem., 154, 414-419, 1986.
9. Newman J M, Lab. Pract., 65-66, 1985.
10. Lengyel Z, Tyihak E, Mincsovics E, J. Liq. Chromatog., 5, 1541-1553, 1982.
11. Porra R J, Eur. J. Biochem., 156, 111-121, 1986.
12. King S Y P, Fung H L, J. Chromatog., 343, 129-137, 1985.
13. Figueira M A, Ribeiro J A, J. Chromatog., 325, 317-322, 1985.
14. Juguelin H, Cassangne C, Anal. Biochem., 142, 329-335, 1984.
15. Gattaveccia E, Tonelli D, Budini R, Radiochem. Radioanal. Lett., 59, 121-130, 1983.
16. Reske S N, Fuchs R, Machulla H J, Winkler C, Radiochem. Radioanal. Lett., 55, 257-264, 1983.
17. Saito M, Kasai N, Yu R K, Anal. Biochem., 148, 54-58, 1985.
18. Schmidt K, Baer H P, Anal. Biochem., 141, 499-502, 1984.
19. Romer J, Wagner H, J. Lab. Comp. Radiopharm., 24, 904- 908, 1987.
20. Wilson I D, Lafont R, Insect Biochem., 16, 33-40, 1986.
21. 'Chromatography of Technetium 99m Radiopharmaceuticals-a Practical Guide', Robbins P J, Soc. Nucl. Med.
22. Bish R E , Silverstein D, Bede J I, J. Radioanal. Chem., 57, 565-573, 1980.
23. Filthuth H, Synth. Appl. Isotop. Lab. Comp. Proc. 2nd Int. Symp., 1985, ed. Muccino R R, Elsevier, 1986, p465- 472.
24. Bonner W M, Laskey R A, Eur. J. Biochem., 46, 83-88, 1974.
25. Hammermaier A, Reich E, Bogl W, Nuc-Compact, 16, 200-206, 1985.
26. Grand J, Lab. Pract., 27-29, 1987.
27. Shulman S D, Kobavashi Y, Synth. Appl. Isotop. Lab. Comp. Proc. 2nd Int. Symp. 1985, ed. Muccino R R, Elsevier, 1986, p459-464.
28. Shulman S D, J. Liq. Chromatog. 6, 35-53, 1983.
29. Miwa M, Matsumoto M, Tezuka M, Okada S, Ohsuka S, Fujiwake H, Anal. Biochem., 152, 391-395, 1986.

30. Anisimov Yu S, Cherenko S P, Ivanov A B Kalinin V N, Peshekhonov
    V D, Senchenkov E P, Tyapkin I A, Zanevsky Yu V, J. Chromatog.
    178, 117-124, 1979.
31. Anisimov Yu S, Abduskurov D A, Dao C C, Tchan C D, Cheremukhina G
    A, Cherenko S P, Hafner K, Ivanov A B, Movchan S A, Netushil T,
    Peshekhonov V D, Smykov L P, Zanevsky Yu V, Nucl. Instr. Meth.
    Phys. Res., B17, 524- 526, 1986.
32. 'The Radiochromatography of Labelled Compounds': Review 14,
    Sheppard G, The Radiochemical Centre, Amersham, p10-20, 1972.
33. Still W C, Kahn M, Mitra A, J. Org. Chem., 43, 2923-292, 5,
    1978.
34. Hays S J, J. Lab. Comp. Radiopharm., 24, 351-360, 1987.
35. Senderoff S G, Heys J R, Blackburn D W, J. Lab. Comp.
    Radiopharm., 24, 971-978, 1987.
36. 'High Performance Liquid Chromatography', Knox J H, Edinburgh
    University Press, 1982.
37. Scott R P W, J. Chromatog. Library, 33, 1-261, 1986.
38. Nissinen E, Anal. Biochem., 144, 247-252, 1985.
39. Toth L A, Scott M C, Elchisak M A, Life Sci., 39, 519- 529,
    1986.
40. Thaker D R, Boehlert C, Kirk K L, Antkowiak R, Craveling C R, J.
    Biol. Chem., 261, 178-184, 1986.
41. Shafiee A, Trzaskos J M, Paik Y, Gaylor J L, J. Lipid Res., 27,
    1-10, 1986.
42. Lafont R, Beydon P, Blais C, Garcia M, Lachaise F, Riera F, Somme
    G, Girault J P, Insect Biochem., 16, 1-6, 1986.
43. Honda A, Morrison A R, McCluski E R, Needleman P, Biochim.
    Biophys. Acta, 794, 403-410, 1984.
44. Eling T E, Danilowicz R M, Henke D C, Sivarajah K, Yankaskas J R,
    Boucher R C, J. Biol. Chem., 261, 12841- 12849, 1986.
45. Henke D, Danilowicz R M, Eling T E, Biochim. Biophys. Acta, 876,
    271-279, 1986.
46. Schlesinger D H, 'C.R.C. Handbook of Hplc Separation of Amino
    Acids, Peptides and Proteins', ed. Hancock W, C.R.C. Press,
    1984, p367-378.
47. Lentzen H, Simon R, J. Chromatog., 389, 444-449, 1987.
48. Lohmander L S, Anal. Biochem., 154, 75-84, 1986.
49. Bonn G, J. Chromatog., 387, 393-398, 1987.
50. Svoboda M, Lambert M, Moroder L, Christophe J, J. Chromatog.,
    296, 199-211, 1984.
51. Linde S, Welinder B S, Hansen B, Sonne O, J. Chromatog., 369,
    327-339, 1986.
52. Hnatowich D J, Nucl. Med. Biol., 13, 373-377, 1986.
53. Sabouni A H, Ma J K H, Malanga C J, Prep. Biochem., 16, 259-272,
    1986.
54. Boothe T E, Finn R D, Vora M M, Emran A M, Kothari P J, Wukovnig
    S, Synth. Appl. Isotop. Lab. Comp. Proc 2nd Int. Symp. 1985, ed.
    Muccino R R, Elsevier, 1986, p453-458.

55. Bell T D, Synth. Appl. Isotop. Lab. Comp. Proc. 1st Int. Symp., 1982, eds. Duncan W P, Susan A B, Elsevier, 1983, p217-222.
56. Wilkinson D J, Lockley W J S, J. Lab. Comp. Radiopharm., 22, 883-892, 1985.
57. Wilkinson D J, Lockley W J S, J. Lab. Comp. Radiopharm. in press.
58. Drenth B F H, De Zeeuw R A, Int. J. Appl. Radiat. Isotop., 33, 681-683, 1982.
59. Podowski A A, Feroz M, Mertens P, Khan M A Q, Bull. Environ. Contam. Tox., 32, 301-309, 1984.
60. Meyer G J, Osterholz A, Hundeshagen H, J. Radioanal. Chem., 80, 229-235, 1983.
61. Shiue C-Y, Bai L-Q, Teng R-R, Wolfe A P, J. Lab. Comp. Radiopharm., 24, 55-64, 1987.
62. DeJesus O T, van Moffaert G J , Glock D, Goldberd L I, Friedman A M, J. Lab. Comp. Radiopharm., 23, 919-925, 1986.
63. Satyamurthy N, Bida G T, Luxen A, Barrio J R, J. Lab Comp. Radiopharm., 23, 1045-1046, 1986.
64. Ehrin E, Gawell L, Hogberg T, dePaulis T, Strom P, J. Lab. Comp. Radiopharm., 24, 931-940, 1987.
65. Morecombe D J, J. Chromatog., 389, 389-395, 1987.
66. Takahashi K, Murakami M, Hogami E, Sasaki H, Kondo Y, Misusawa S, Nakamichi H, Ioda H, Miura S, Kanno I, Uemura K, Ido T, J. Lab. Comp. Radiopharm., 23 ,1111-1113, 1986.
67. Dezaro D, Hartwick R A, Chromatog. Sci., 28, 113-136, 1984.
68. Jinno K, Hirata Y, Hiyoshi Y, J. High Res. Chromatog. Chromatog. Commun., 5, 102-103, 1982.
69. Ego D, Beaucourt J P, Pichat L, J. Lab. Comp. Radiopharm., 23, 553-564, 1986.
70. Cundy K C, Crooks P A, J. Chromatog., 281, 17-33, 1983.
71. Clark B J, Fell A F, Chem. Brit., 23, 1069-1071, 1987.
72. Roberts R F, Fields M J, J. Chromatog., 342, 25-33, 1985.
73. Karmen A, Malikin G, Lam S, J. Chromatog., 302, 31-41, 1984.
74. Karmen A, Malikin G, Freundlich L, Lam S, J. Chromatog., 349, 267-274, 1985.
75. Reeve D R, Crozier A, J. Chromatog., 137, 271-282, 1977.
76. Veltkamp A C, Das H A, Frei R W, Brinkman U A Th, Eur. Chromatog. News, 1, 16-21, 1987.
77. Baba S, Horie M, Watanabe K, J. Chromatog., 244, 57-64, 1982.
78. Baba S, Synth. Appl. Isotop. Lab. Comp. Proc. 2nd Int. Symp., 1985, ed. Muccino R R, Elsevier, 1986, p 479-484.
79. Makey L N, Rodriguez P A, Schroeder F B, J. Chromatog., 208, 1-8, 1981.
80. Kessler M J, see reference 2, Chapter 7, 150-170, 1986.
81. 'Modern Practice of Gas Chromatography', 2nd Ed.,ed. Grob R L, John Wiley ans Sons Ltd, 1987.
82. 'Gas Chromatography', Willet J E, John Wiley and Sons Ltd, 1987.
83. Kokes R I, Tobin H, Emmet P H, J. Amer. Chem. Soc., 77, 5860-5862, 1955.

84. Matucha M, Smolkova E, J. Chromatog., 127, 163-201, 1976.
85. Cram S P, Advances in Chromatography, Vol. 9, ed. Giddings J C, Keller R A, p 243-293, 1970.
86. Herkner K, Schaupal R, Goedl U, J. High Res. Chromatog. Chromatog. Commun., 8, 558-561, 1985.
87. Jones J R, Poncipe C, Barton J M, Wright W W, Brit. Polym. J., 18, 312-315, 1986.
88. Reissmann P, Eckert H, Eckert G, Nahrung, 30, 679-686, 1986.
89. Emmanuel B, Fenton T W, Turner B V, Milligan L P, J. Biochem. Biophys. Meth., 8, 1-7, 1983.
90. Carrol L, Jones J R, Shore P R, J. Lab. Comp. Radiopharm., 24, 763-772, 1987.
91. Campbell I M, Anal. Chem., 51, 1012A-1021A, 1979.
92. Doerfler D L, Emmons G T, Campbell I M, Anal. Chem., 54, 832-833, 1982.
93. Markey S P, Abramson F P, Anal. Chem., 54, 2375-2376, 1982.
94. Dick R M, Freeman J J, Kosh J W, J. Chromatog., 347, 387- 392, 1985.
95. Lee C R, Esnaud H, Pollit R J, J. Chromatog., 387, 505- 508, 1987.
96. Karmen A, Longo N S, J. Chromatog., 112, 637-642, 1975.
97. Cori O, Diaz E, Lab. Pract., 35, 78-79, 1986.
98. Rodriguez P A, Culbertson C R, Eddy C L, J. Chromatog., 264, 393-404, 1983.
99. Herkner K, Chromatographia, 16, 39-43, 1982.
100. Baba S, Akira K, Sasaki Y, Furuta T, J. Chromatog., 382, 31-38, 1986.
101. Long M A, Carroll L, Jones J R, Tang Y S, J. Chromatog., 287, 381-384, 1984.
102. Pellon G, Michel G, Anal. Lett., 19, 1511-1521, 1986.
103. Dunn M J, J. Chromatog., 418, 145-185, 1987.
104. Smith C L, Cantor C R, Nature, 319, 701-702, 1986.
105. Abou-Zeid C, Smith I, Grange J, Steele J, Rook G, J. Gen. Microbiol., 132, 3047-3053, 1986.
106. Garoff H, Ansorge W, Anal. Biochem., 115, 450-457, 1981. 107. Roberts P L, Anal. Biochem., 147, 521-524, 1985.
108. Bishop C W, Kendrick N C, Santek D A, Thompson R G, DeLuca H F, Anal. Biochem., 148, 133-148, 1985.
109. Kaniansky D, Rajec P, Svec A, Havasi P, Macasek F, J. Chromatog., 258, 238-243, 1983.
110. Leseticky L, Radioisotopy, 26, 113-126, 1985.
111. Klein P D, Adv. Chromatog., 3, 3-65, 1966.
112. Wilzbach K M, Riesz P, Science, 126, 748-749, 1957.
113. Gordon B E, Otvos J W, Erwin W R, Lemmon R M, Int. J. Appl. Radiat. Isotop., 33, 721-724, 1982.
114. Chizhov V P, Sinitzina, L A, J. Chromatog., 104, 327- 335, 1975.
115. Bigeleisen J, J. Chem. Phys., 34, 1485-1493, 1961.
116. Jansco G, Van Hook W A, Chem. Rev., 74, 689-750, 1974.

117. Keogh T, Strife R J, Rodriguez P A, Sanders R A, J. Chromatog., 312, 450-455, 1984.
118. Van Hook W A, J. Chromatog., 328, 333-336, 1985.
119. Gerardy B M, Poupaert J H, Vendervorst D, Dumont P, Declerq J-P, Van Meersshe M, Portoghese P S, J. Lab. Comp. Radiopharm., 22, 5-22, 1985.
120. Heck H D A, Simon R L, Anbar M, J. Chromatog., 133, 281- 290, 1977.
121. Filer C N, Fazio R, Ahern D G, J. Org. Chem., 46, 3344- 3346, 1981.
122. Kaspersen F M, Funke C W, Sperling E M G, van Rooy F A M, Wagenaars G N, J. Chem. Soc. Perkin Trans. II, 585- 591, 1986.
123. Powell W S, Anal. Biochem., 128, 93-103, 1983.
124. Powell W S, Adv. Prostag. Thrombox. Leuko. Res., 11, 207-213, 1983.
125. Baweja R, J. Liq. Chromatog., 9, 2609-2621, 1986.
126. Baweja R, J. Chromatog. 369, 125-131, 1986.
127. Tanaka N, Thornton E R, J. Amer. Chem. Soc., 96, 1617- 1619, 1976.
128. Yeung P K F, Hubbard J W, Baker B W, Looker M R, Midha K K, J. Chromatog., 303, 412-416, 1984.

4
Modern Spectrometric Methods for the Analysis of Labelled Compounds

F.M. Kaspersen, C.W. Funke, G.N. Wagenaars, and P.L. Jacobs

SCIENTIFIC DEVELOPMENT GROUP, ORGANON INTERNATIONAL B.V., P.O. BOX 20, 5340 BH OSS, THE NETHERLANDS

1 INTRODUCTION

A proper analysis of chemical compounds should give information about the chemical identity (not only the structure but also enantiomeric form), the chemical purity and chemical composition (e.g. giving information about counter-ions, solvents of crystallization). For labelled compounds information is also needed about isotopic purity (defined as the % of isotope present in the compound), the position/distribution of the isotope in the molecule and degree of labelling/specific activity (see Table 1). In earlier days such information was not obtained easily partly because of the lack of sensitivity of analytical methods for labelled compounds of which usually small amounts are available and partly because of lack of specific detection techniques. For instance, tritiated molecules were analyzed using radiochromatography and troublesome degradation reactions. However, in the past ten years the possibilities for spectrometric analyses of labelled compounds have increased enormously and this chapter will give an overview of these methods with the exception of (radio)chromatography that will be dealt with in another chapter[2].

Table 1: Characteristics of labelled compounds.

	Preferred method of analysis
Chemical identity	NMR; MS; IR
Enantiomeric purity	chromatography; NMR
Chemical purity	chromatography; NMR
Isotopic purity	chromatography; NMR
Position of isotope	NMR
Distribution of isotope	NMR
Degree of labelling	MS; UV

2 NMR SPECTROSCOPY

It is not exaggerated to state that modern NMR spectroscopy is one of the most powerful techniques for the analysis of organic compounds. The availability of high-field instruments and the development of special pulse sequences give the spectroscopist valuable information concerning structure and conformation of chemical products. In this section we will deal with the applications of the different NMR techniques for the analysis of labelled compounds. The NMR properties of the different isotopes are summarized in Table 2 while for the theoretical background of modern pulse techniques recent reviews are available[3,4].

Table 2: NMR-properties of some commonly used isotopes.

Isotope	Spin*	Relative sensitivity	NMR-frequency (MHz at 4,7 Tesla)	Natural abundance (%)
^1H	$\frac{1}{2}$	$\equiv 1$	200	99,98
^2H	1	$9,65 \times 10^{-3}$	30,7	$1,5 \times 10^{-2}$
^3H	$\frac{1}{2}$	1,21	213,3	0
^{13}C	$\frac{1}{2}$	$1,59 \times 10^{-2}$	50,3	1,108
^{14}C	0	NMR inactive		0
^{15}N	$\frac{1}{2}$	$1,04 \times 10^{-3}$	20,3	0,37
^{17}O	$\frac{5}{2}$	$2,91 \times 10^{-2}$	27,1	$3,7 \times 10^{-2}$
^{18}O	0	NMR inactive		2,04

*Nuclei with spins ≥ 1 give rise to broadening of signals due to quadrupole relaxation.

^1H NMR spectroscopy

The most obvious method for the analysis of labelled compounds – with the exception of ^3H material (see below) – is ^1H NMR spectroscopy. Not only is confirmation of chemical structure obtained but also information can be obtained concerning position/amount/ distribution of isotopic substitution. The information about the label is possible:1) in the case of deuterated compounds by measurement of the decrease of intensity of signals in the ^1H NMR spectrum, 2) by a change in the patterns of the signals by introduction of extra couplings (^{13}C, ^{15}N-labelled compounds) or change in coupling constants (^2H compounds) and 3) by shifts of the signals of protons in the vicinity of isotopes.

For deuterated compounds the easiest method for determination of the distribution and amount of label is the measurement of the decrease of the integral of the signals. However, one should realize that the precision of this measurement – even under ideal nOe-suppressed conditions – is only around 5% and especially low amounts of deuterium (which can be measured but not quantified by ^2H NMR) are easily missed.

Since ^{2}H–^{1}H couplings are only 15% of the values of the corresponding ^{1}H–^{1}H couplings (see Table 3), information can also be gained about e.g. stereochemistry from the remaining coupling patterns. In Figure 1 an example is given for [6–^{2}H]–androstenedione 2 prepared by solvolysis of the enol ether 1; the change of the doublet (J=2,5 Hz) at 5,75 ppm of H(4) to a singlet indicates that the deuterium at position 6 is introduced at the β–position[5] in agreement with earlier results by Malhotra and Ringold[6] obtained with IR spectroscopy.

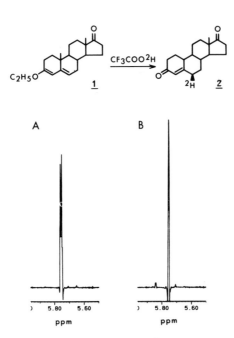

Figure 1: ^{1}H NMR spectra (360 MHz; in C^{2}HCl$_3$) of H(4) of androstenedione (2) A) unlabelled B) [6β–^{2}H]–androstenedione

For ^{13}C labelled compounds information about position/amount of isotope can be obtained by the ^{13}C–^{1}H coupling constants (see Table 3). An example is given in Figure 2 for [aniline–^{13}C]–bepridil (3). As in the case of ^{2}H the amount of ^{13}C can be calculated from the ratio coupled/uncoupled signal, however, the precision of such a measurement is strongly dependent on the complexity of the signal. For ^{15}N labelled compounds we have about the same situation as ^{13}C (see Table 3) but at the same time broadening of protons directly bound to nitrogens (because of the quadrupole of ^{14}N) disappears resulting in sharp/split signals.

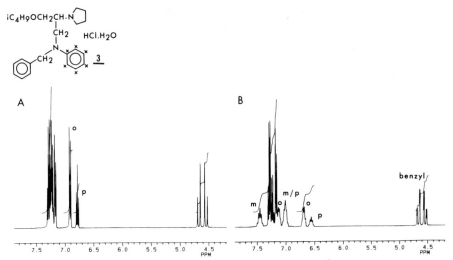

Figure 2: Part of the ^{1}H NMR spectrum (360 MHz; in C^{2}HCl$_3$) of
bepridil·HCl·H$_2$O (3) A) unlabelled B) [^{13}C$_6$–aniline]–bepridil; 99% ^{13}C

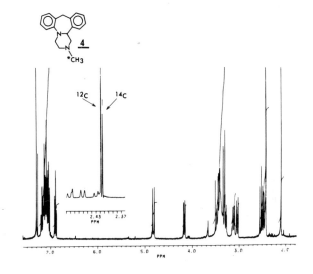

Figure 3: ^{1}H NMR (360 MHz; in C^{2}HCl$_3$) spectrum of a mixture of
[N–methyl–^{14}C]–mianserin and natural abundance mianserin (4).

   Introduction of heavier isotopes introduces chemical shifts. Since
these effects are rather small (for ^{2}H approx. 0,01 ppm[7], for [37]Cl
approx. −0,2 ppb[8]) they are of limited diagnostic value except in cases
where one is dealing with isotopes that are NMR inactive[9] such as ^{14}C or
[18]O. An example for ^{14}C is given in Figure 3 in which the NMR-spectrum
of a mixture of [N-methyl-^{14}C]-mianserin (4) and natural abundance-
mianserin is shown; in this case the isotopic shift is −3,8 ppb.

   The information obtained from ^{1}H NMR spectra for tritiated
material of high specific activity is of limited value. Since one is
dealing with small amounts of material (25-100 μg) the ^{1}H spectra are
often of poor quality because of low signal/noise ratios and the
interference by plasticizers/ paraffinic oil present in the solvents
used for purification/storage of the labelled compound. As is
illustrated in Figure 4 for Org 3770 (5) the specific activity for this
compound measured on basis of this [1]H-spectrum can be anything between 0
and 30 Ci/mmol while the real value (as determined by mass spectrometry)
is 21 Ci/mmol. The ^{1}H spectra are in general of more value to determine
specific impurities in the labelled compound such as catalysts for the
tritiation-reaction[10] or buffers that have been used during
HPLC-purifications.

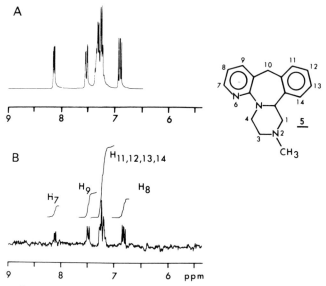

Figure 4: ^{1}H NMR spectra (200 MHz; in deuterated DMSO) of Org 3770 (5).
A) unlabelled compound B) [12-^{3}H]-Org 3770

   Although the [1]H-^{3}H coupling is 6% larger than the corresponding
[1]H-^{1}H coupling, the difference is in the order of the precision of the
measurement and thus hardly measurable. A different situation is met

when by introduction of tritium the local symmetry is disturbed such as in methyl groups or such as in tritiated $NaBH_4$; in the ^1H NMR spectrum $Na^{11}BH_3{}^3H$ is visible as doublets, $Na^{11}BH_2{}^3H_2$ as triplets and $Na^{11}BH^3H_3$ as quartets[11] (J_{3H-1H}= 11,1 Hz).

Table 3: Heteronuclear coupling constants (Hz).

$^1H-^2H$	:	15% of corresponding $^1H-^1H$ coupling
$^1H-^3H$	:	107% of corresponding $^1H-^1H$ coupling
$^1H-^{13}C$	:	one bond: 5,7 (%s character bond) – 18,4 Hz[12]; two bonds: –5 to +2 Hz; three bonds: up to 11 Hz
$^1H-^{15}N$	:	one bond: 60-90 Hz (negative)[13]; two bonds: 0-2 Hz (sp^3), up to 20 Hz (sp^2); three bonds: up to 7 Hz

^2H-NMR spectroscopy

Since the shielding (which determines the chemical shift of a nucleus) of deuterium is almost completely determined by the local environment in a molecule, the chemical shifts in ^1H, ^2H (and ^3H) spectroscopy are identical. As a consequence a direct comparison between ^1H and ^2H spectra of the labelled compound and thus assignment of the ^2H signals is possible. However, deuterium NMR spectroscopy suffers from low sensitivity and, because of the quadrupole of deuterium, broadening of signals. While for small molecules still relatively sharp signals are observed – e.g. in the small cyclopentanes distinction is possible between mono-, di- and multideuterated forms (upfield shifts 7 up to 53 ppb)[14] – for larger molecules broadening is a serious problem. For instance, we could not observe the deuterium signal for the peptide H–Met(O_2)–Glu–His–Phe–D-Lys–Phe–OH (6) deuterated in the phenylalanine ring. Because of this broadening $^1H-^2H$ couplings are hardly visible as illustrated for 7 in Figure 5; in contrast $^{13}C-^2H$ couplings because of their larger values, are still of diagnostic value in [$^2H-^{13}C$]-doubly labelled compounds.[15]

Despite its limitations ^2H-NMR is still the technique of choice for determination of the position of deuterium especially for compounds containing a small fraction of deuterium which is not visible by other NMR techniques. For instance in [14b-^2H]-mianserin (4b), prepared by total synthesis, the small amount (10%) of ^2H present in position 1 could not be observed with ^1H NMR, but was easily detected by ^2H NMR[16].

Figure 6

Two-dimensional techniques have also been applied for the analysis of deuterated compounds. Gould et al.[17] applied the original pulse sequence of Maudsley and Ernst[18] to measure the $^2H-^1H$ correlation shift of the bis-acetonide of [4-^2H]-allose (8) (Figure 7). By the correlation of the broad ^2H-signal with the sharp proton signals more information should be obtainable from the ^2H NMR

spectra; however, problems are still expected with respect to sensitivity and resolution for deuterated molecules of high molecular weight.

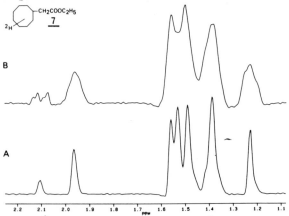

Figure 5: ^{2}H NMR spectra (55 MHz in C^{2}HCl$_3$ at 50 $^\circ$C) of the ethyl ester
of cyclooctylacetic acid (7). A) ^{1}H-decoupled B) ^{1}H-coupled

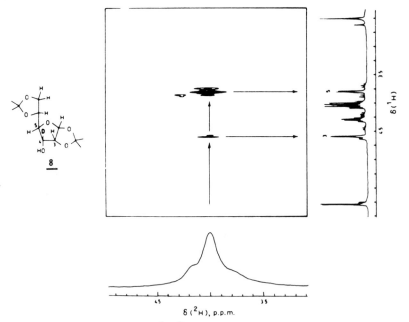

Figure 7: Contour plot of ^{1}H/^{2}H COSY experiment of the bisacetonide of
[4-^{2}H] allose (8).

(Reproduced from reference 18)

^{3}H NMR spectroscopy

Tritium NMR is now an established technique for the analysis of tritiated compounds[19]. While in the first publications large amounts were needed – e.g. for the analysis of [^{3}H-ethyl]–ethylbenzene 10 Ci was used[20] –, it is possible to obtain satisfactory S/N ratios using 1–5 mCi amounts. As is shown in Figure 8, it is even possible to achieve such a quality that the ^{13}C–satellites (0,55% intensity of the main peak) are visible.

With ^{3}H NMR information can be obtained about:
1. Chemical identity/radiochemical purity
2. Position of tritium in the molecule
3. Distribution of the label and distribution of labelled species
4. Specific activity
5. Enantiomeric purity

In general the information concerning chemical identity/ radiochemical purity present in ^{3}H NMR spectra is limited. Like in ^{2}H NMR the ^{3}H chemical shift can be compared directly to the corresponding ^{1}H–chemical shifts. Only small deviations are possible e.g. when the ^{1}H spectrum and ^{3}H spectrum are recorded under different conditions (e.g. in ^{2}H$_2$O–solutions ^{1}H–decoupling can introduce shifts as a result of heating of the sample). Furthermore,[21,22] the ω_r/ω_d ratio is dependent on the hybridization of the C–H bond[21,22] but resulting deviations are at maximum 0,1 ppm[22]. Thirdly, the secondary isotope shifts (in ditritiated material) range from ~0,03 ppm for geminally ditritiated material to ~0,01 ppm for vicinally ditritiated material and <0,01 ppm for "allylic" ditritiated material[23]. Because of the high concentrations of radioactivity in the samples radiolysis in the NMR tube is a complication, especially when deuterated chloroform is used as NMR solvent since it might result in false negative interpretations of the ^{3}H NMR spectra. By radiolysis deuterated HCl is formed which leads to exchange of tritium at labile positions[24], the formation of protonated forms of amines (and subsequent shifting of signals to lower field) and isomerization reactions. In Figure 9 an example of such a radiation induced reaction for Org OD 14 (10) is shown. To minimize the possible effects of radiolysis we always add carrier to the NMR samples.

Identification of unknown labelled compounds by ^{3}H NMR is difficult since only the tritiated part of the molecule is visible. One of the few examples we encountered is given in Figure 10; in this case the side product 13 formed in the synthesis of tritiated 12 was easily identified on the basis of the vinylic(7,5 ppm) and benzylic (4,6 ppm) signals in the ^{3}H NMR.

Of course, the main application of ^{3}H NMR is the determination of the position of tritium in a labelled compound on the basis of the chemical shift and/or ^{1}H–^{3}H–, and ^{3}H–^{3}H coupling constants. Numerous examples are now available in the literature; especially the unexpected scrambling of tritium in molecules is easily detected such as the

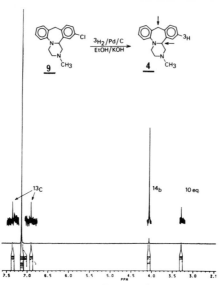

<u>Figure 8</u>: ^{3}H NMR (360 Mz; in C^{2}HCl$_3$; ^{1}H-decoupled) of [13-^{3}H]-mianserin (<u>4</u>) prepared by reductive dechlorination of 13-chloromianserin (<u>9</u>).

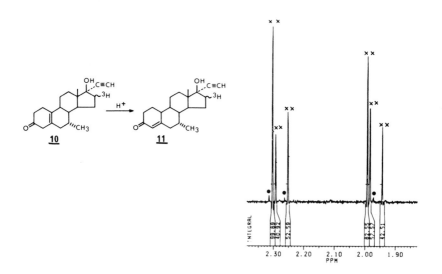

<u>Figure 9</u>: ^{3}H NMR spectra (384 Mz; in C^{2}HCl$_3$) of [16-^{3}H]-Org OD 14 (<u>10</u>). ●: (<u>10</u>)   **: (<u>11</u>)

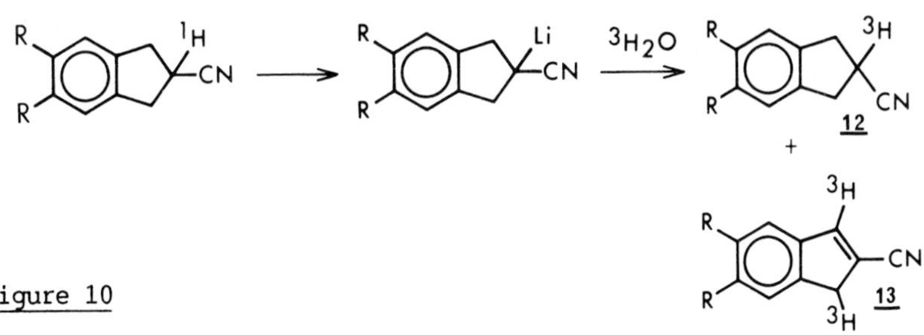

Figure 10

vinylic/allylic exchange on labelling of the locust adipokinetic hormone (15) by reduction with $^3H_2$ of the 4,5–dehydroleucine–precursor (14)[25] or the general labelling of cyclooctylacetic acid (17) by reduction with $^3H_2$ of cyclooctenylacetic acid (16)[26] (Figure 11).

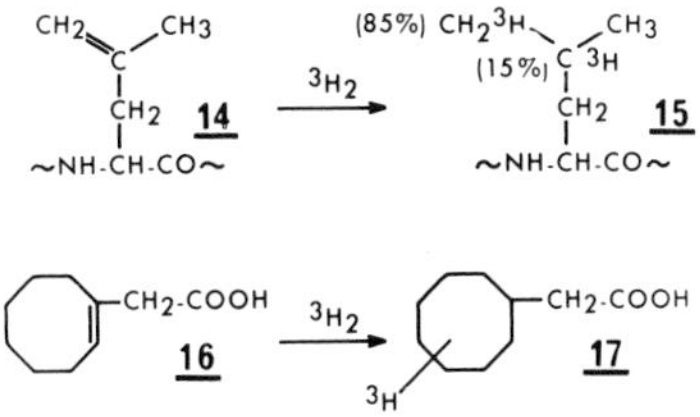

Figure 11

In cases where the chemical shift is not conclusive ^1H–^3H couplings might be helpful; for instance the tritium introduced at position 6 in cyclo-pentylether of oestradiol (18) by catalytic exchange with $^3H_2$ could be assigned at the 6β position on the basis of the vicinal ^1H–^3H coupling of 12Hz[24] (Figure 12).

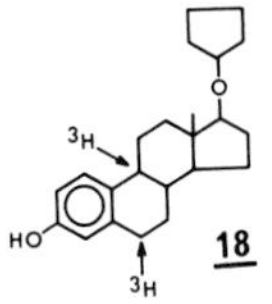

Figure 12

A topic that is hardly discussed in the literature is the detection limit in ^3H NMR. As is illustrated for [13-^3H]–mianserin in Figure 8 this is a function of the S/N ratio. While at low S/N ratios the compound seems to be labelled exclusively at position 13, the high quality spectrum (shown in Figure 8) indicates substantial incorporation at 14b (3,8%) and 10 eq. (0,5%). Since most spectra published do not reach such S/N ratios, one should be careful in defining the limits of isotopomeric purity.

Measurement of the <u>distribution</u> of mono– and multi–<u>labelled</u> <u>molecules</u> is only possible when we are dealing with vicinally, geminally and sometimes allylic multitritiated compounds. For multitritiated material ^{3}H–^{3}H couplings and small upfield isotopic shifts in comparison with the monotritiated material are present. Examples are given in Figure 13a and b for Org 3236 (<u>19</u>) and tritiated [^{2}H]–DMSO (<u>20</u>) respectively (in the latter case the isotopic shift is caused by ^{2}H).

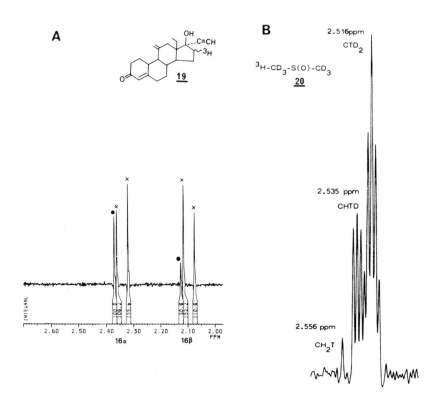

<u>Figure 13</u>: A: ^{3}H NMR spectrum (384 MHz in C^{2}HCl$_3$; nOe–suppressed) of Org 3236 (<u>19</u>); ●: monotritiated   x: ditritiated
B: ^{3}H NMR spectrum (384 MHz in ^{2}H–DMSO) of ^{3}H–deuterated DMSO (<u>20</u>)

^{3}H NMR specta are generally recorded using ^{1}H broad–band decoupling resulting in a ^{3}H signal intensity enhancement (nOe) which may differ considerably for differently labelled molecules. Therefore, precise ratio measurements have to be done under nOe–suppressed

conditions[27], especially for ratios between mono and geminally
ditritiated molecules or for compounds in which the tritium atoms are
located in proton-poor regions e.g. steroids 21 and 22[28] respectively
(Figure 14).

Figure 14

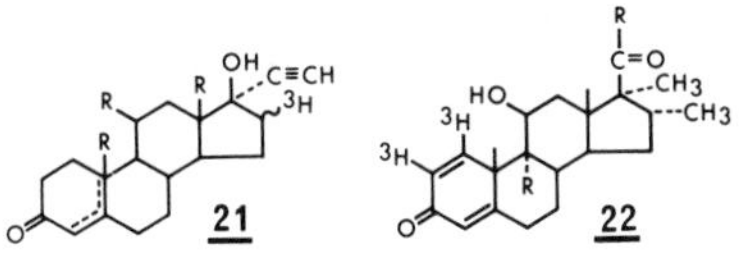

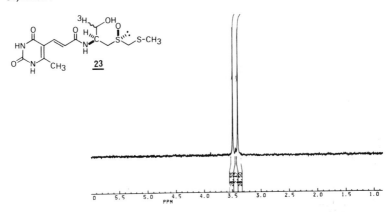

In principle the specific activity could be derived from the
ratio mono-tritiated/multititriated material if one assumes that no
unlabelled product is present but in our experience this is hardly
ever the case. For this reason it is better to use mass spectrometry
for the determination of the specific activity. A special case is with
steroids of type 21 which were labelled by exchange of the 17-oxo-
precursors with tritiated water followed by ethynylation[23]. Assuming
that there are no isotope effects operative and that there is no
preference for the α- or β-side of the steroid one can calculate the
specific activity with the formula:

specific activity= 57,52 $[1-(2R+1)^{-1}]$ Ci/mmol

with R being the ratio di/monotritiated material. For instance for 19
we calculated a value of 46,0 Ci/mmol while mass spectrometry gave
47,4 Ci/mmol.

Figure 15: ^{3}H-NMR spectrum (384 MHz, in [2]H-DMSO, [1]H-decoupled) of
[3]H-sparsomycin (23).[29]

In some compounds the analysis of the tritium NMR spectrum is complicated by the fact that by introduction of tritium a pair of diastereoisomers is formed because the tritium distorts the symmetry of the molecule. An example is given in Figure 15 for tritiated sparsomycine (23); the tritium signals of monotritiated material show up as two singlets while the corresponding signal of the proton in the ^{1}H NMR spectrum is an AB-system. As observed for bepridil tritiated in the pyrrolidine ring (3) even more complex spectra were obtained due to a mixture of two monotritiated diastereoisomers and cis and trans ditritiated material.[30]

Enantiomeric purity can also be measured with ^{3}H NMR. While Elvidge et al.[31] applied tris(3-hepta-fluorobutyryl-d-camphorato)-europium (III) for the determination of secondary alcohols, we used Pirkle's alcohol (1-phenyl-2-trifluoroethanol (24)) for the measurement of the enantiomeric purity/ absolute configuration for a series of aliphatic amines[30,32]. The problem with such measurements is that the tritium is not always located at those positions which show maximum splitting. This is illustrated in Figure 16 where the ^{3}H-NMR spectrum of racemic [10-^{3}H]-Org 3770 (5) in the presence of Pirkle's alcohol is shown; whereas the signal for [3]H_{eq} shows the expected 1:1 splitting, the signal for [3]H_{ax} is not resolved.

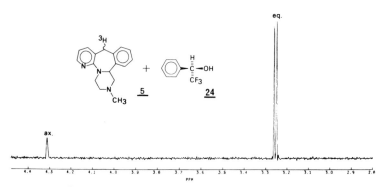

Figure 16: [3]H NMR spectrum (384 MHz, in C^2HCl_3, ^{1}H decoupled) of [10-^{3}H]-Org 3770 (5) in the presence of Pirkle's alcohol (24).

Two-dimensional techniques have been applied in ^{3}H NMR. Sergent and Beaucourt[33] applied [3]H-^{3}H correlation spectroscopy for the interpretation of the (by the existence of different rotameric forms) very complicated ^{3}H-spectrum of PK-11195 (25) (Figure 17). We investigated the possibilities of both [3]H-^{3}H and [3]H-^{1}H-correlation NMR spectroscopy[34] (see Figure 18). These studies indicated that for most applications [3]H-^{3}H correlation spectra could be best recorded using the original pulse sequence developed by Maudsley and Ernst[18] because of the variation in the [1]H-^{3}H coupling constants. The detection limit for such measurements is about 3 mCi per labelled position resulting in NMR samples with large amounts of radioactivity.

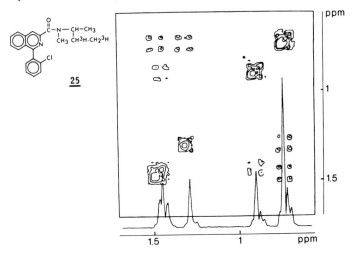

Figure 17: ³H-³H-COSY-45 SPECTRUM (106 MHz, ¹H-decoupled) of <u>25</u>.
Reproduced from reference 33 by permission of Pergamon Press.

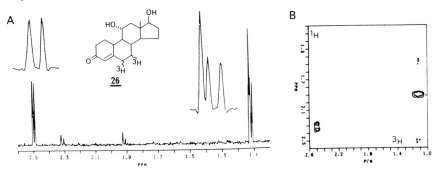

Figure 18: ³H NMR spectra (384 MHz; solvent C²HCl₃) of <u>26</u>
A) one-dimensional ³H spectrum (¹H decoupled) B) ³H-¹H correlation
plot using the pulse sequence of Maudsley and Ernst[18]
Reproduced from reference 34 by permission of John Wiley and Sons.

¹³C NMR spectroscopy

Of course the obvious technique for the analysis of ¹³C-labelled
compounds is ¹³C-NMR. Since the natural abundance of ¹³C is only 1,1%,
the labelled positions are clearly visible. The chemical shift gives
unambiguous information on the position of the isotope, while the
so-called DEPT-technique reveals the number of protons attached to the
carbon-13 atom. The percentage label can be measured from the increase

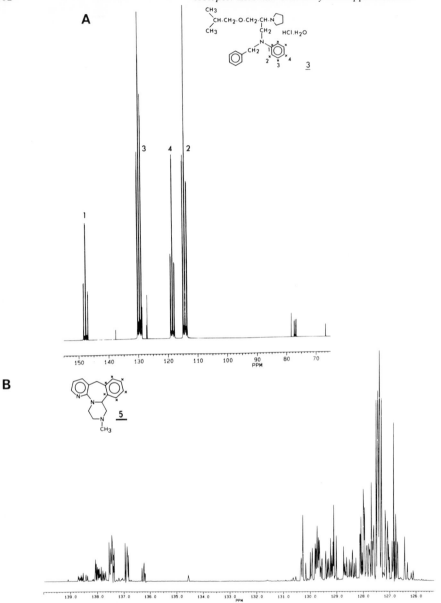

Figure 19: A: 90 MHz ^{13}C-spectrum of [aniline-^{13}C]-bepridil·HCl·H$_2$O (3); solvent C^{2}HCl$_3$, ^{1}H decoupled; ^{13}C content 99% B: 90 MHz ^{13}C-spectrum of [benzene-^{13}C]-Org 3770 (5); solvent C^{2}HCl$_3$, ^{1}H decoupled; ^{13}C-content 99%.

in intensity of the ^{13}C–signals in comparison to the unlabelled compound but not in a simple way because the intensity of ^{13}C–signals is strongly dependent on the number of hydrogens attached. In the case of multiple–labelled ^{13}C–compounds the spectra can be complicated by ^{13}C–^{13}C couplings (see Figure 19). While for [aniline–^{13}C]–bepridil (3) rather simple spectra are obtained (Figure 19a), the resonances in the spectrum of ^{13}C–Org 3770 (5) (Figure 19b) are complicated by second–order effects due to the almost identical chemical shifts of both the protonated and the non–protonated aromatic carbon atoms.

^{13}C NMR spectroscopy is also of great use for the analysis of deuterated compounds. Carbon atoms that are directly attached to deuterium atoms undergo an upfield shift of 0,3–0,6 ppm/^{2}H and show ^{2}H–^{13}C coupling. Moreover the intensity of the deuterated ^{13}C–signals decreases because of decreased nOe, slow relaxation and broadening by the quadrupole moment of deuterium. For the carbon atoms that are two or three bonds away from the deuterium a small upfield shift (0,01–0,1 ppm/^{2}H)[35] is observed but there have also been reports for zero or negative shifts. As an example of such effects the ^{13}C–spectrum of [6β,16α,16β–^{2}H$_3$]–danazole (27) is given in Figure 20. Although ^{3}H gives also an isotopic shift in ^{13}C NMR[36] this is of limited diagnostic value in view of the low sensitivity of ^{13}C and subsequent large amounts of radioactivity needed.

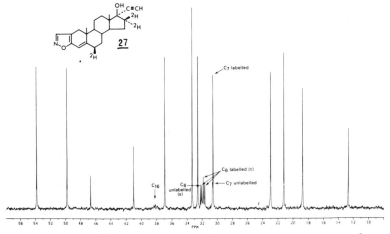

Figure 20: 90 MHz ^{13}C NMR spectrum (^{1}H decoupled; solvent (C^{2}HCl$_3$) of [6β,16α,16β–^{2}H$_3$]–danazole (27). According to ^{2}H NMR it contains >95% ^{2}H at 16β and 16α and ~90% ^{2}H at 6β.

With the INEPT (Insensitive Nuclei Enhanced by Polarisation Transfer)[37,38] pulse sequence we can selectively detect ^{13}C atoms that are directly attached to ^{2}H atoms. As shown in Figure 21 the sensitivity is greatly enhanced. Tandem–SEFT (J modulation of ^{13}C–spin echo signals under both ^{1}H and ^{2}H decoupling) distinguishes the ^{13}C NMR signals of quaternary carbons, CH, CH_2, CH_3, CH^2H, $CH_2{}^2H$, CH^2H_2, C^2H, C^2H_2 and C^2H_3-groups[39]. An example for the test mixture 29 (Table 4) is given in Figure 22. Although these techniques cannot be applied on routine NMR spectrometers, they may give useful information for e.g. doubly labelled compounds; in general mass spectrometry is a good alternative.

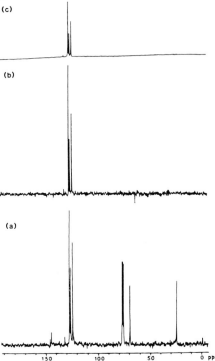

Figure 21: ^{13}C NMR spectra of 1-phenylethanol (28) (in $C_1^2HCl_3$).
a) normal ^{1}H decoupled spectrum of 5% phenylethanol   b) ^{13}C {^{2}H} INEPT of 5% [benzene-$^2H_5$]-phenylethanol in 1% C^2HCl_3, 240 transients c) 1% [benzene-$^2H_5$]-phenylethanol, 20000 transients

^{13}C NMR spectroscopy also contains valuable information for ^{14}C labelled compounds of high specific activity. As with other heavy isotopes the introduction of ^{14}C causes small upfield shifts for the neighbouring C-atoms (-20 ppb)[40]. More important information is contained in the intensity of the ^{13}C-signals. Since ^{14}C (produced by the ^{14}N (n,p) ^{13}C reaction) contains no ^{13}C, the signals for the labelled C-atoms decrease or are absent (of course the same holds for [^{12}C]-compounds). Altman et al.[40] used this method for the analyses of ^{14}C-carbohydrates while we applied the method for the determination

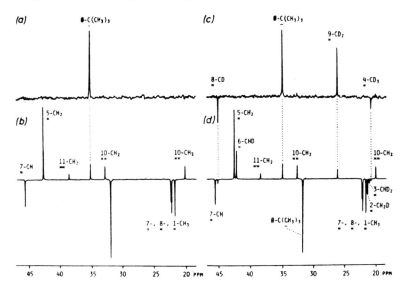

Figure 22: ^{13}C NMR Tandem SEFT spectra of the mixture $\underline{29}$ (100,6 MHz in C^2HCl_3).

Reproduced from reference 39 by permission of the American Chemical Society.

Table 4: Composition of test mixture ($\underline{29}$) with partially deuterated alkyl groups (C_6F_6, δ=17,0 ppm, internal lock).

Compound	δ	Compound	δ
1. $C_6H_5CH_3$	21,80	7. $(C_6H_5)_2CHCH_3$	45,54; 22,43
2. $C_6H_5CH_2D$	21,53	8. $(C_6H_5)_2CDCH_3$	45,11; 22,32
3. $C_6H_5CHD_2$	21,25	9. $(CD_2)_6$	26,40
4. $C_6H_5CD_3$	20,97	10. $(CH_3CH_2CH_2CH_2)_2O$	20,24; 32,82
5. $(C_6H_5)_2CH_2$	42,62	11. $C_6H_5CH_2CH_2C_6H_5$	38,52
6. $(C_6H_5)_2CHD$	42,27	12. t-butylbenzene	31,92 (CH_3); 35,10 (C)

of ^{14}C in Org 3770 ($\underline{5}$)[9]. We used this method also for the determination of the specific activity of ^{14}C-mianserin ($\underline{31}$) prepared by reaction of ^{14}C-methyl iodide with N(2)-demethylmianserin ($\underline{30}$)[41] (Figure 23).

Figure 23:

Both ^{18}O and ^{15}N can also be detected by ^{13}C NMR spectroscopy on the basis of the upfield shift on the neighbouring ^{13}C-atom (e.g. for ^{18}O 0,03 ppm)[42] while ^{15}N can also be detected on the basis of the $^{13}C-^{15}N$ coupling constants.

NMR spectroscopy of other nuclei

For compounds labelled with other NMR-active isotopes such as ^{15}N or ^{17}O it is of course possible to use ^{15}N or ^{17}O NMR for the analysis of these compounds. However, if we are dealing with compounds which contain high percentages of these isotopes, it may be more convenient to use ^1H NMR and ^{13}C NMR spectroscopy or mass spectrometry for the assessment of the amount of isotope and chemical identification. Only if there is a possibility for exchange/rearrangement reactions may the NMR techniques give extra information.

3 MASS SPECTROMETRY

Mass spectrometry is a very useful technique for the analysis of organic compounds and the information obtained is often complementary to NMR. The molecular weight and molecular composition (exact mass) as well as structural information (fragmentation reactions) can in general be obtained with this technique. For isotopically labelled compounds the amount of isotope and the distribution of labelled species can also be measured while the location of the label can be determined by studying suitable fragmentation reactions[43,44]. For the latter measurement NMR is always better suited when scrambling of the label during fragmentation occurs; in fact, labelling (mainly with ^2H) is often applied to study the scrambling reactions. The percentage of label can be measured with very high precision (<0,001%, at least for compounds of high chemical purity) by conversion of the molecule in small gaseous fragments (e.g. for ^{13}C and ^{18}O analyses CO_2) and determination of isotope ratios in specially designed mass spectrometers; however, such precision is only required for special purposes (e.g. age determinations) and usually straightforward analysis of (quasi-) molecular ions or fragment-ions is sufficient.

Until ten years ago mass spectra were usually recorded using the electron-impact (EI)-mode or chemical ionization (CI)-mode. Both ionization techniques suffer from the fact that the sample has to be evaporated and therefore thermally unstable or very polar samples cannot be analyzed this way unless troublesome derivatisation is applied. With both techniques fragmentation reactions can occur and these fragmentation reactions can be influenced by isotopic substitution. For instance in the CI-spectrum (neg. ion mode) of the the trimethylsilyl derivative of Lormetazepam (32; Figure 24)[45] the fragment ion [M-HCl-TMSOH]$^{-\cdot}$ is the base peak for unlabelled material at temperatures above 210 °C, whereas for the deuterated compound the [M-HCl]$^{-\cdot}$ ion is more abundant because of the more difficult elimination of TMS-O^{2}H.

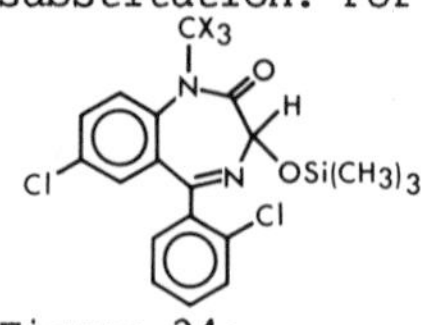

X=$^1H,^2H$

32

Figure 24:

With Field Desorption (FD)- and Fast Atom Bombardment (FAB) mass spectrometry the samples can be analyzed without evaporation and these techniques – in both positive- or negative-ion mode – are now commonly applied methods. Even polar and unstable compounds give in general abundant (quasi)-molecular ions while fragmentation-reactions are limited in comparison with EI. Numerous examples can be found in literature[46-49] both for stable and radioactive isotopes. In this review two examples from our own work are given (Figures 25 and 26); both compounds do not show any molecular ions under EI- or CI- conditions.

It is very difficult to make a statement about the sample size needed for a mass spectrum because this is highly dependent on the spectrometer, the ionization method used and the nature of the sample. With direct introduction techniques usually 0,1 μg–1 μg is needed, an amount that should be no problem for a synthetically prepared isotopically labelled compound except when very short-lived isotopes are applied.

By applying so-called two-dimensional techniques such as GC-MS or LC-MS sensitivity and selectivity can be improved, while for GC-MS the sample size can be reduced to 0,01–10 ng. In the first case only volatile samples[51,52] (which can be reached by derivatization[53]) can be measured while in the latter case the choice of volatile HPLC-mobile phases can limit the possibilities of this technique. Another very promising technique is MS-MS[53] in which selected (molecular) ions obtained with one of the ionization techniques can be studied in more detail by collision-induced dissociation. The possibilities of this method for labelled compounds are illustrated in the study of Mallis et al.[54] of ^{18}O labelled nucleotides.

The isotopic distribution in a sample of a labelled compound can be determined using a straightforward calculation method after correction for natural abundances[55]. The isotopic content of a compound can also be determined by minimizing the differences between the calculated pattern (for an estimated isotopic content) and the experimental determined pattern. An example of such an exercise is given in Table 5 for the tritiated peptide Org 2766 (6). In general these calculations are done with mass spectra obtained with unit resolution but it is also possible to use average masses of complex isotope clusters[56]. When mass spectra show strong [M-H]-peaks the calculations can fail for deuterated compounds in case the [M-H] peak is composed of [M-H] and [M-^{2}H] peaks of unknown ratio; special calculations have been developed for cases like this[57,58]. As an alternative the (EI) spectra can be recorded at reduced electron energy since the [M-H]-effect is often energy dependent[55] or derivatives that do not show the [M-H]-ions can be prepared.

For FAB and FD measurements the precision of the calculated isotope contents is between 1-5% (if better precision is needed this can be reached by construction of calibration curves of mixtures of labelled and unlabelled species[59]), whereas the detection limit for

H.L.Met(O₂)-L.Glu-L.His-L.[³H]-Phe-D-Lys-L-Phe-OH
6

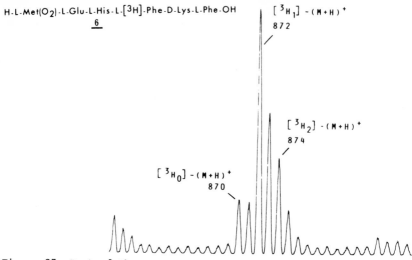

Figure 25: Part of the FAB-mass spectrum (positive ion-mode; thioglycerol as matrix) of [³H]-Org 2766 (**6**) prepared as described in [50]
Reproduced from reference 50 by permission of John Wiley and Sons.

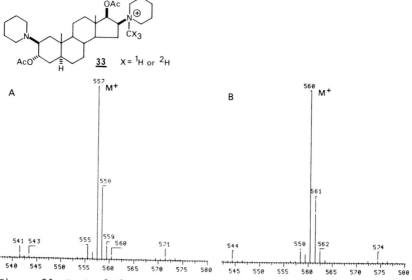

Figure 26: Part of the FAB-spectrum (positive ion-mode) of Org NC 45
(**33**); glycerol as matrix.
A: unlabelled Org NC 45   B: [N-methyl-²H₃]-Org NC 45

<u>Table 5</u>: Calculation of the specific activity of tritiated Org 2766 (6) (see Figure 25). All intensities are given in %.

a)Experimental isotopic pattern used as basis for the quantitative evaluation
b)Quantitative evaluation of the ion pattern given in Figure 25 with respect to the relative contribution of the nonlabelled, the singly, and doubly tritiated species of Org 2766 and comparison of the reconstructed and the experimental ion abundance pattern. (rel.int.: relative intensity; expt.: experimental; recon.: reconstructed; diff.: difference)
c)Calculation of the specific activity of Org 2766.

a)
species	(M+H-2)	(M+H-1)	(M+H)	(M+H+1)	(M+H+2)	(M+H+3)
rel.int (%)	5.6	7.4	100	57.5	23.3	6.4

b)
m/z	rel.int. expt.(%)	$[^3H_0]$	$[^3H_1]$	$[^3H_2]$	recon. pattern	diff.(%) recon.-expt.
868	2.1	0.9	-	-	0.9	- 1.1
869	2.9	1.2	-	-	1.2	- 1.7
870	20.9	15.6	5.4	-	21.0	+ 0.1
871	20.1	9.0	7.1	-	16.1	- 4.0
872	100	3.6	95.5	0.9	100	0
873	56.8	1.0	54.9	1.3	57.2	+ 0.4
874	38.9	-	22.3	16.5	38.8	- 0.1
875	16.6	-	6.2	9.5	15.7	- 0.9
876	5.0	-	1.1	3.8	4.9	- 0.1

c)
species	rel.abundance(%)	rel.specific activity
$[^3H_0]$	12.3	0
$[^3H_1]$	75.1	799.1
$[^3H_2]$	12.6	268.2

specific activity 1,067.3 TBq mmol^{-1}

the measurement of the isotopic content is between 1 and 5%[60]; this means that these values can be measured with a precision of 5%. For lower specific activities UV-measurements (see below) are more appropriate.

Unlike in NMR artefacts are not uncommon in mass spectrometry, especially under FAB conditions[60,61] and they can interfere with specific activity determinations. For instance with thioglycerol as matrix irreproducible reduction of the S-S bond in (tritiated) peptides of the vasopressin type was observed[61]. Sometimes irreproducible [M+H]$^+$-ions are also formed. Such an example is shown in Table 6 for Org 6632 (34): under FD-conditions strong and non-reproducible [M+H]$^+$-peaks were present making it impossible to calculate the specific activity on the basis of the M-cluster. Fortunately the [M-HF]$^+$·-cluster behaved normally and this could be used to calculate the specific activity (which was in good agreement with the values measured by other methods).

Table 6: FD-spectrum of Org 6632.

M—cluster			M—HF$^{]+\cdot}$—cluster		
m/z	Exp. int. (%)	Theor. int (%)	m/z	Exp. int. (%)	Theor. int. (%)
406	6,5		388	100	100
407	14,5		389	26,7	26,3
408	100	100	390	38,8	36,3
409	85,3	26,3	391	14,6	9
410	40,3	36,3			
411	46,7	9			
412	12,9	1,3			
413	4,8	0,1			

At increased ion source pressure the intensity of the $[M+H]^+$—ions under EI—conditions will increase proportionally due to ion—molecule reactions. Long residence times in the ion source will also enhance the probability of ion—molecule reactions, so it is recommended to operate at a high repeller voltage to reduce the residence time. Care should be taken to measure the labelled and unlabelled samples under identical instrumental conditions preferably during the same session.

Especially for molecules containing high amounts of deuterium or tritium fractionation during evaporation, during GC—analysis[62] or by introduction from a reference inlet may occur, resulting in a variation of the mass spectrum with time. However, by signal accumulation during the total ionization curve or by construction of calibration curves of mixtures of the labelled and unlabelled compound this problem can easily be circumvented.

Finally, exchange of label during MS—analysis has been reported[63], even for compounds labelled in a stable position. For instance replacement of trideuteromethyl groups in quaternary amines by methyl groups using methane as reagent gas in CI mass spectrometry has been reported[64]; no loss of label was observed with ammonia as reagent gas.

4 IR SPECTROSCOPY

IR is a very useful technique for the organic chemist because, in contrast to NMR, it provides direct information about functional groups (e.g. hydroxyl, carbonyl, nitro groups) while it is the best spectroscopic technique to determine the crystalline form of solids. The application of the Fourier—Transform technique increased the sensitivity of IR enormously (only a few μg are needed) while new sampling techniques (DRIFT, PAS)[65] broadened the possibilities for sample-forms and dimensions and the coupling of IR with GC made it possible to study more complicated mixtures.

In IR we are observing molecular vibrations and the frequencies are directly related to the mass of the vibrating atoms according to:

$$\nu(cm^{-1}) = \infty \sqrt{\frac{f}{\mu}}$$

$$\mu = m_A \cdot m_B / (m_A + m_B)$$

Hence substitution of one element by its heavier isotope will lead to a shift of the absorption band to lower wave numbers. As expected from this formula the isotope shifts for heavier elements like carbon, nitrogen and oxygen are small, and moreover are generally

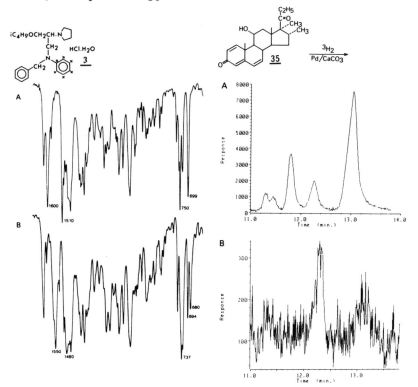

Figure 27: IR-spectra (KBr disc) of bepridil·HCl·H$_2$O (A) and [aniline-^{13}C]-bepridil·HCl·H$_2$O (B).

Figure 28: GC/IR chromatogram of the reaction mixture on reduction of (35) with ^{2}H$_2$. A: 1697–1705 cm^{-1} (C=O) B: 2140–2170 cm^{-1} (C–^{2}H)

overlapped by the rest of the spectrum. This is illustrated in Figure 27 where the IR-spectra of bepridil·HCl·H$_2$O (3) of natural abundance and [aniline-^{13}C]-bepridil·HCl·H$_2$O (3a) are shown. There are clear differences but these are not simply related to the isotope position and content, so the diagnostic value of these shifts is limited. This was also demonstrated by the IR-spectrum of [aniline-^{14}C] bepridil·HCl·H$_2$O with a specific activity of 10 mCi·mmol^{-1}; the spectrum was completely identical to the spectrum of the compound of natural abundance.

C-^{2}H bonds are an exception because C-^{2}H vibrations are located in a quiet region of the IR-spectrum (2000–2200 cm^{-1}). As illustrated in Figure 28 even with this free frequency the diagnostic value is limited because the intensity of a C-^{2}H band is very low resulting in poor S/N ratios.

Although it is possible to determine the stereochemistry of deuterium from the position of the C-^{2}H band[6,66], NMR (both ^{2}H and ^{1}H) should be preferred because of their higher sensitivity.

5 UV SPECTROSCOPY

With UV spectroscopy we are measuring electronic transitions and since isotopic substitution only mildly disturbs the electronic density of a compound very small effects are expected. This is demonstrated in Figure 29 for benzene labelled in different positions[67]; no information can be obtained from these spectra regarding the type, the amount and the position of the isotopes.

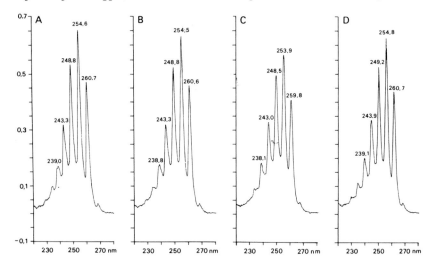

Figure 29: UV spectra of benzene in cyclohexane. A: natural abundance benzene; B: [^{2}H$_1$]-benzene; C: [^{2}H$_6$]-benzene; D: [^{13}C$_6$]-benzene

Under special conditions deuterium can influence phosphorescence[68] but these effects are of no relevance to the analysis of isotopically labelled compounds. Besides being a mass detector in HPLC, UV-spectra are mainly applied in the determination of the specific activity of radioactive compounds; the radioactivity is measured by counting while the concentration is determined on basis of the UV absorption of the labelled material. With compounds of high chemical purity such as ^{14}C-products a direct UV spectrum is possible but in case of tritiated material a prepurification is often necessary. As is illustrated in Figure 30 HPLC combined with a photo-diode-array detector is an ideal combination.

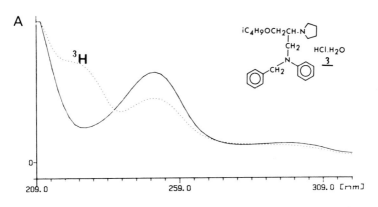

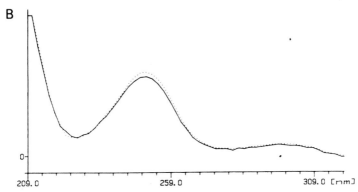

Figure 30: UV spectra of bepridil (3) A: unlabelled bepridil (–) and [pyrrolidine-³H]-bepridil of 97% radiochemical purity (...); spec. activity 20 Ci/mmol without HPLC; B: ³H-bepridil after HPLC (µBondapak C_{18} with 0,01M aq. NH_4OAc/acetonitrile; photo-diode-array-detector: HP 1040A)

6 MISCELLANEOUS TECHNIQUES

As mentioned in the introduction, chromatography can be used for the determination of (radio)chemical purity and chemical identity. By combination with a flow-through counter radioactivity can be analyzed. However, it is also possible to detect ^{14}C-labelled compounds by combination of GC with a microwave discharge interface which converts the compound to CO_2; the $^{14}CO_2$ is selectively detected by a conventional mass spectrometer[69]. For deuterated molecules it is also possible to measure the atomic emission spectra using a microwave discharge detector[70]. For both methods the spectroscopic methods are applied as mass detectors while the actual analysis of the labelled compound is done on the basis of retention time.

7 CONCLUDING REMARKS

As was illustrated in this overview there are nowadays a large number of spectrometric methods available for the analysis of labelled compounds. The selection of a technique depends mainly on the percentage of label that is present in the compound. For high amounts (>50%) MS and both direct NMR methods (where the isotope itself is measured) and indirect NMR (measurement by isotope effects) methods can be applied. For intermediate isotopic content (5% <>50%) MS is still a good choice but of the NMR methods only the direct methods are useful. For low isotopic contents (<5%), which is only interesting for radioactive compounds, we have to rely on UV/radioactive measurements instead of MS for the measurement of the specific activity while for tritiated material ^{3}H NMR still gives useful information (at least when enough material is available). For ^{14}C-compounds of low specific activity we have to rely on chromatographic techniques only.

For future developments there is still a need for increasing sensitivity and selectivity; especially for radioactive compounds with shorter half-lives such as ^{35}S only chromatography and biological assays are available. A very promising development could be MS-MS while new NMR pulse-sequences (rather than higher fields) may contribute to more selective analyses.

8 ACKNOWLEDGEMENT

The authors thank Fons van Rooy, Irene Schlachter, Eric Sperling and Kitty Willemsen for their contributions to the preparations of the labelled compounds. Howard Berkeley is acknowledged for his critical reading of the manuscript and Angelie de Kadt for her assistance in preparation of the manuscript.

9 REFERENCES

1. E.A. Evans, Tritium and Its Compounds, 2nd ed. Butterworth, London, 1974.

2. W.J.S. Lockley, The Radiochromatography of Labelled Compounds, previous chapter of this book.
3. R. Benn and H. Günther, Angew. Chem. Int. Ed. Eng., 1983, 22, 350.
4. G.A. Morris, Magn. Res. Chem., 1986, 24, 371.
5. M. Groen, J. Leemhuis, A. Broess and F.M. Kaspersen, unpublished results.
6. S.K. Malhotra and H.J. Ringold, J. Am. Chem. Soc., 1965, 87, 3228.
7. V.R. Haddon and L.M. Jackman, Org. Magn. Reson., 1973, 5, 333.
8. T. Schaefer and R. Sebastian, J. Amer. Chem. Soc., 1987, 109, 6508.
9. C.W. Funke, F.M. Kaspersen, E.M.G. Sperling and G.N. Wagenaars, J. Chem. Soc. Chem. Commun., 1986, 462.
10. C.N. Filer in Isotopes in the Physical and Biomedical Sciences, Labelled Compounds (Part A), E. Buncel and J.R. Jones (Ed.), Elsevier, 1987, Vol. 1, p. 186.
11. J.L. Altman and L. Thomas, Anal. Chem., 1980, 52, 992.
12. F.W. Wehrli and T. Wirthlin, Interpretation of Carbon-13 NMR Spectra, Heyden, 1976.
13. G.C. Levy and R.L. Lichter, Nitrogen-15 Nuclear Magnetic Resonance Spectroscopy, John Wiley and Sons, 1979.
14. A.C. Faro, C. Kemball, R. Brown and I.H. Sadler, J. Chem. Res. (S), 1982, 342.
15. A.P.W. Bradshaw, J.R. Hanson, R. Nyfeler and I.H. Sadler, J. Chem. Soc. Chem. Commun., 1981, 649.
16. F.M. Kaspersen, J.S. Favier, G.N. Wagenaars, J. Wallaart and C.W. Funke, Recl. Trav. Chim. Pays-Bas, 1983, 102, 457.
17. S.J. Gould, V.A. Palaniswamy, H. Bleich and J. Wilde, J. Chem. Soc. Chem. Commun., 1984, 1075.
18. A.A. Maudsley and R.R. Ernst, Chem. Phys. Lett., 1977, 50, 368.
19. E.A. Evans, D.C. Warrell, J.A. Elvidge and J.R. Jones, Handbook of Tritium NMR Spectroscopy and Applications, J. Wiley and Sons, 1985.
20. G.V.D. Tiers, C.A. Brown, R.A. Jackson and T.N. Lahr, J. Am. Chem. Soc., 1964, 86, 2526.
21. J.M.A. Al-Rawi, J.P. Bloxsidge, C. O'Brien, D.E. Caddy, J.A. Elvidge, J.R. Jones and E.A. Evans, J. Chem. Soc. Perkin Trans. 2, 1974, 1635.
22. J.P. Bloxsidge, J.A. Elvidge, J.R. Jones, R.B. Mane and M. Saljoughian, Org. Magn. Reson., 1979, 12, 574.
23. F.M. Kaspersen, C.W. Funke and G.N. Wagenaars, unpublished results.
24. C.W. Funke, F.M. Kaspersen, J. Wallaart and G.N. Wagenaars, J. Labelled Compd. Radiopharm., 1983, 20, 843.
25. D.E. Brundish in Synthesis and Applications of Isotopically Labeled Compounds, R.R. Muccino (Ed.), Elsevier, Amsterdam, 1986, p. 365.
26. F.M. Kaspersen, I.H.G. Schlachter, C.W. Funke, G.N. Wagenaars and J.R. Mellema, to be published.
27. D. Shaw, Fourier Transform NMR Spectroscopy, Elsevier, Amsterdam 1976, p. 290.
28. F.M. Kaspersen, C.W. Funke, E.M.G. Sperling and G.N. Wagenaars, J. Labelled Compd. Radiopharm., 1987, 24, 219.
29. F.M. Kaspersen, I.H.G. Schlachter and L. van de Broek: unpublished results.

30. F.M. Kaspersen, C.W. Funke, F.M.G. Sperling, F.M. van Rooy and G.N. Wagenaars, J. Chem. Soc. Perkin Trans. 2, 1986, 585.

31. J.A. Elvidge, E.A. Evans, J.R. Jones and L.M. Zhang, Synthesis and Applications of Isotopically Labeled Compounds, R.R. Muccino, (Ed.), Elsevier, Amsterdam, 1985, p. 401.

32. F.M. Kaspersen, C.W. Funke and G.N. Wagenaars in Synthesis and Applications of Isotopically Labeled Compounds, R.R. Muccino (Ed.), Elsevier, Amsterdam, 1985, p. 355.

33. L. Sergent and J.P Beaucourt, Tetrahedron Lett., 1985, 5291.

34. C.W. Funke, G.N. Wagenaars and F.M. Kaspersen, Magn. Reson. Chem., 1986, 24, 434.

35. D.A. Forsyth in Isotopes in Organic Chemistry, Isotopic Effects, E. Buncel and C.C. Lee (Ed.), Elsevier , 1984, Vol. 6, p. 1.

36. C.H. Arrowsmith, L. Baltzer, A.J. Kresge, M.F. Powell and Y.S. Tang, J. Am. Chem. Soc., 1986, 108, 1356.

37. D.M. Doddrell, J. Staunton and E.D. Laue, J. Chem. Soc. Chem. Commun., 1983, 602.

38. P.L. Rinaldi and N.J. Baldwin, J. Am. Chem. Soc., 1982, 104, 5791.

39. J.R. Wesener, P. Schmitt and H. Günther, J. Am. Chem. Soc., 1984, 106, 10.

40. L.J. Altman, R.E. O'Brien, S.K. Gupta and H.R. Schulten, Carbohydr. Res., 1980, 87 189.

41. F.M. Kaspersen, C.W. Funke and G.N. Wagenaars, unpublished results.

42. E. Ponnusamy, C.R. Jones and D. Fiat, J. Labelled Compd. Radiopharm. 1987, 24, 773.

43. J.H. Beynon, Mass Spectrometry and its Applications to Organic Chemistry, Elsevier, Amsterdam, 1960.

44. M. Kiessling, W. Schwarz and R. Tümmler, Radiochem. Radioanal. Lett., 1982, 52, 65.

45. S. Takahashi, Biomed. Environm. Mass Spectrom., 1987, 14, 257.

46. U. Bahr and H.R. Schulten, J. Labelled Compd. Radiopharm., 1979, 18, 571.

47. H.R. Schulten and W.D. Lehmann, Biomed. Mass Spectrom., 1980, 7, 468.

48. W.D. Lehmann and M. Kessler, Fresenius Z. Anal. Chem., 1982, 312, 311.

49. W.D. Lehmann and F.M. Kaspersen, J. Labelled Compd. Radiopharm., 1984, 21, 455.

50. F.M. Kaspersen, A.M. van Rooy and J. Wallaart, Recl. Trav. Chim. Pays-Bas, 1983, 102, 450.

51. H. Kanamaru, R. Takai, M. Horiba, I. Nakatsuka and A. Yoshitake, Radioisotopes (Jap.), 1985, 34, 67.

52. H. Kanamaru, I. Nakatsuka and A. Yoshitake, Radioisotopes (Jap.), 1985, 34, 401.

53. S. Lewis and M. Yudkoff, Anal. Biochem., 1985, 145, 354.

54. L.M. Mallis, F.M. Raushel and D.H. Russell, Anal. Chem., 1987, 59, 980.

55. B.J. Millard, Quantitative Mass Spectrometry, Heyden, London, 1978, p. 60.

56. K. Blom, C. Dybowski, B. Munson, B. Gates and L. Hasselbring, Anal. Chem., 1987, 59, 1372.

57. W. Benz, Anal. Chem., 1980, 52, 248.
58. K. Blom, J. Schuhardt and B. Munson, Anal. Chem., 1985, 57, 1986.
59. R.H. Liu, F.P. Smith, I.A. Low, E.G. Piotrowski, W.C. Damert, J.G. Phillips and J.Y. Liu, Biomed. Mass. Spectrom., 1985, 12, 638.
60. W.D. Lehmann in Synthesis and Applications of Isotopically Labeled Compounds, R.R. Muccino (Ed.), Elsevier, Amsterdam, 1985, p. 293.
61. W.D. Lehmann and F.M. Kaspersen, unpublished work.
62. Y. Cherrah, J.B. Falconnet, M. Desage, J.L. Brazier, R. Zini and J.P. Tillement, Biomed. Environm. Mass Spectrom., 1987, 14, 653.
63. Reference 55, p. 68.
64. G.C. Ford, S.J. Grigson and N.J. Haskins, Biomed. Mass Spectrom., 1976, 3, 230.
65. P.R. Griffith and M.P Fuller in "FT-Infrared Spectrometry of Powdered Samples", in Advances in Infrared and Raman Spectroscopy, R.J.H. Clark and R.E. Hester (Eds), Heyden, London, 1981, Vol. 9, Chap. 2.
66. S.K. Malhorta and H.J. Ringold, J. Am. Chem. Soc., 1964, 86, 1997.
67. Using laser photon-resonance MS the difference in absorption - maximum between ^{12}C-benzene and [mono-^{13}C]-benzene in the gas phase has been determined as 0,02 nm (Bruker Report 2/1987, p. 5).
68. N. Kanamura, H.R. Bhattacharjee and E.C. Lim, Chem. Phys. Lett., 1974, 26, 174.
69. S.P. Markey and F.P. Abramson, Anal. Chem., 1982, 54, 2375.
70. H.G. Hege and J. Weymann, Anal. Chem. Symp. Ser., 1982, 11, 679.

5
Localization and Quantitation of Radioactivity in Solid Specimens Using Autoradiography

M.A. Williams

DEPARTMENT OF ANATOMY AND CELL BIOLOGY, THE UNIVERSITY, SHEFFIELD, S10 2TN, UK

1 INTRODUCTION

The detection of natural or experimentally placed radioactivity in solid specimens is carried out against a
backdrop of other structural studies. That is to say that
autoradiography is largely used to answer questions
concerned with how radioactivity is dispersed among the
distinguishable components of a material. This contrasts
with other measurement methods (perhaps gamma camera
studies apart) which are primarily concerned with 'how
much' rather than 'where'. Autoradiography is thus almost
unique in being concerned with spatial localization – in
being an 'anatomical' approach. Indeed, only if localization is a major objective is autoradiography likely to
be the way forward.

The term autoradiography (and its now largely out-
moded synonym radioautography) is applied to a body of
methods applicable in a wide range of fields. In several
of these fields its use is popular, routine even, yet
often the practitioners are unaware of the manner in which
the approach is employed elsewhere, and they will often
assume that if autoradiography is mentioned then what is
implied is their own familiar routine. It may well be
that it is not. It might be that what is meant is some
quite different anatomical level of work or some quite
different degree of refinement in image processing and
interpretation. Autoradiography covers a multiplicity
of scales, tracer isotopes, detector materials, fields of
application and types of question posed. Autoradiography
might be defined in an effort to cover all these
possibilities as: a collection of techniques whereby
radioactivity is localized within solid specimens by means

of the close apposition of a detector layer. An account of autoradiography might possibly cover its history, mechanism, methodology, materials, laboratory conditions, quantitation, fields of application, place alongside other experimental approaches, ultimate limits, achievements so far or future path of development. Alternatively, it may be directed to one particular aspect of method, one segment of possibiities or one important field of application. By its place this account has to cover theory, mechanism, place alongisde other experimental approaches and possibilities. In the course of dealing with these matters, some account will necessarily be given of materials and their handling and storage, certain areas of application, limits and possibilities of autoradio-graphy, experimental design and control experiments, details of test systems and standards, and choice of radiolabels and chemicals. It will no doubt be appreciated that readers requiring more explicit detail about particular topics should consult appropriate specialist volumes.[1,2,3,4,5]

2 RADIOISOTOPES FOR AUTORADIOGRAPHIC USE

The great majority of work has been carried out with β-emitters such as hydrogen-3, carbon-14, or sulphur-35 indeed more than 90% of the published work involves tritium or carbon-14, principally the former. In some earlier studies isotopes either of higher energy (e.g. phosphorus-32) and/or with attendant gamma ray emissions (e.g. iron-59, iodine-131) found use, but chromatograms apart, these isotopes are now largely out of favour. The recent trend has been towards low energy β-emitters or electron capture (EC) isotopes yielding extremely low energy Auger electrons. Examples of commonly used Auger electron emitters are chromium-51, iron-55 and iodine-125. Indium-111 has also found occasional uses. (See table 1.)

Generally the gamma emissions are of no use for imaging purposes in this field and present, when using iodine-131 for instance, merely an inconvenience, though of course situations can be found where the gamma rays make parallel counter-based procedures more valuable by their possibilities for non-destructive sample counting. X-rays put out by EC isotopes make no significant contribution to the image from specimens coated with thin emulsion layers, though they do contribute when imaging with thicker ones. Using iodine-125 for example, $0.5\mu m$ sections coated with monocrystalline layers of Ilford L4 are imaged very largely via Auger electrons whilst sections in the $5-50\mu m$ range applied to Betamax sheet film

Table 1   Abbreviated data concerning some radioisotopes
           exploited by autoradiography

Element	Isotope mass	Decay type	Half life	Decay energy Max	Infinite* Thickness in d=1.1
Hydrogen	3	Beta	12.26 y	0.018 MeV	5 μm
Carbon	14	Beta	5760 y	0.155 MeV	100 μm
Sulphur	35	Beta	87.2 days	0.167 MeV	-
Iron	59	Beta, gamma	45 days	0.27, 0.46 MeV etc	-
Iron	55	EC	2.7 y	Auger electrons X rays	-
Iodine	131	Beta, gamma	8.04 days	0.61 MeV etc	-
Iodine	125	EC	60 days	Auger, X rays	20 μm
Calcium	45	Beta	165 days	0.254 MeV	-

*Estimated by autoradiography using the dipping method.

(see table 2) give images which have a large contribution
from the x-ray output of the isotope. Specimens
containing low levels of the EC isotopes chromium-51 or
iodine-125 are not hazardous to handle, but blocks of
material containing millicurie quantities are quite
dangerous and must be handled with appropriate
precautions. Solid specimens are very frequently
sectioned on microtomes before study. The sectioning
process requires the operator to sit near the specimen
block for periods of some hours and hence, if millicurie
quantities are present, necessitates encasing the microtome
in suitable shielding material such as lead-doped perspex
(e.g. Premac, Premise Engineering, Watford, U.K.).

Alpha particle emitters produce images composed of
individual straight tracks. These have both advantages
and disadvantages from an interpretational point of view.
However, the main factor limiting their use is the atomic
number of the elements involved. This is very high, and
whilst the elements concerned may feature in geological
material and tissues contaminated with reactor fuels, they
do not make up part of normal healthy living cells and

Table 2   Some nuclear emulsions sold for autoradiography

Manufacturer	Products for microautoradiography			
	Product	Sensitivity	Form	Crystal size
Ilford	L	4	pellciles	0.14μm
Ilford	K	0,2,5	pellicles	0.20μm
Ilford	G	5	pellicles	0.27μm
Eastman-Kodak	129.01	(2)	gel	0.06μm
Eastman-Kodak	NTB	(2)	gel	0.29μm
Eastman-Kodak	NTB 2	(3)	gel	0.26μm
Eastman-Kodak	NTB 3	(5)	gel	0.34μm
Kodak U.K.	AR 10	(2)	stripping plate	0.20μm

	Products for macro or semi micro work			
CEA Verken	Ultrofilm		cuttable sheet	
Amersham International	Hyperfilm 3H		cuttable sheet	
Amersham International	Hyperfilm Betamax		cuttable sheet	

Other emulsions are available from Konishiroku Photo Industry Co., Tokyo and Fuji Film Co., Tokyo. In the U.S.S.R. Niichim Photo, Leningrad also have nuclear emulsions available.

tissues wherein the elements have atomic numbers mostly below forty. Plutonium in bone[6], and polonium, plutonium and uranium-rich dusts in lungs and liver are some of the specialized instances where α-emitters may be objects of study.

At the present time, large numbers of autoradiographic experiments are carried out using tritium (especially in human, animal and plant tissue biology, neurobiology and pharmacology), iodine-125 in studies involving marked proteins and carbon-14 in 'whole body' pharmacokinetic and toxicological studies. These together make up the vast bulk of autoradiography-based work.[7,11]

The emission energies of the electrons used for imaging in these fields of study vary from near zero to 0.155MeV the β-emitters having continuous spectra of energies and the Auger emitters line spectra. The energies employed have important consequences: they partly determine the extent to which the image spreads laterally, ('image spread', 'resolution' - see later and Appendix). Very low energy emitters such as tritium produce electrons which are unable to penetrate right through the sort of section commonly prepared for light microscope examination (3-20μm thickness). The higher energy electrons from carbon-14 or sulphur-35 on the other hand travel far further (perhaps 100μm)[8] resulting in the signal strength of such autoradiographs isotopes being sensitive to specimen thickness i.e. sections containing tritium can sometimes be regarded as 'infinitely thick' whereas those containing carbon-14, sulphur-35 or higher energy β-emitters may not be assumed to be so. Ground sections of bone, rock or hard materials are much thicker than histological sections and if the material is of high density infinite thickness might possibly occur. Table 1 gives details of relevant properties of radioisotopes commonly or occasionally featuring in autoradiographic experiments. See also Figure 1.

3 CHOOSING RADIOCHEMICALS FOR BIOLOGICAL EXPERIMENTS BY AUTORADIOGRAPHY

It has been pointed out elsewhere that not only are the choices of chemical precursor and radioisotope important but also the site of placement of the radioisotope within the tracer molecule.

Care is necessary concerning the placement of tritium, carbon-14 or iodine isotopes, since the results in biological systems often vary with the position of the labelling. This is especially so since the mixtures of radiochemical metabolites that will probably be generated are not being separated and identified (or at least, can only be by parallel experiments).

The most advantageous placements of tritium in the most widely used precursors for RNA, DNA and proteins, are wellknown (uridine 5-H[3], thymidine 6-H[3], and L-leucine 4,5H[3] respectively) and require no further comment. Even when using such well studied and chosen precursors, it should be borne in mind that the usual ('conventional') histological procedures extract water soluble substances, thus removing the unmetabolised precursor and its

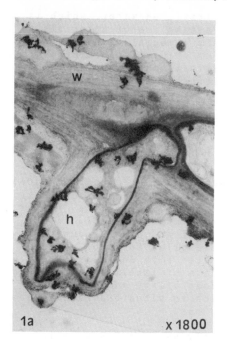

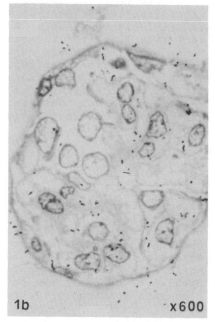

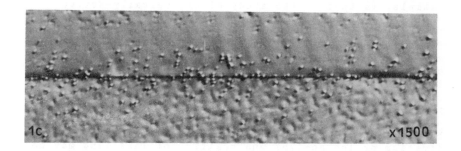

Figure 1   Three autoradiographs.   1a   an electron
microscope autoradiograph using Ilford L4 emulsion
developed in D19 developer.   The material is Tussilago
farfara infected with Puccinia poarum and fed tritiated
glycerol.   Symbols: H haustorium, W host cell wall.   Note
the form of the silver grains.   1b   light microscope
autoradiograph of kidney glomerulus labelled with
$Na_2S^{35}O_4$.   1c A hot line source made to measure image

spread of iodine 125 in recoring $5\,\mu m$ sections, recorded by
Nomarski interference microscopy.

immediate products as well as tritiated water, which is a
major labelled product of the metabolism of materials such
as tritiated thymidine. Readers especially interested in
work with tritiated thymidine or uridine are referred to
texts on cell kinetics or classical texts on tritiated
compounds.[9] Experimenters interested in the soluble
products of such precursors will require autoradiographic
methods based on frozen or freeze-dried specimens. The
relations between those methods and ones employing
conventional liquid fixatives are discussed in a later
section of this account.

Radiochemicals of the type mentioned above are
normally purchased from a supplier and the user is faced
with no requirement to synthesize. This situation is true
for thousands of radiochemicals, the onus on the
experimenter being simply that of choosing well.
Laboratories in the pharmaceutical and other industries
often have to synthesize radiolabelled forms of unique
chemicals and are thus faced often with decisions about
which isotope to use and the site of labelling, decisions
which must take into account feasibility and cost. The
problem of choosing the position to situate a radioactive
atom is a constantly recurring one.

One field which is currently very active wherein
these questions arise and which can therefore serve as an
example, is that of the study of biologically-active
peptides. The isolatable natural peptides are numerous.
In addition many laboratories create biologically-active
fragments or other derivatives. Study of each species of
molecule needs the synthesis of a radiolabelled form with
consequent choices of radioisotope and position of
labelling. Debates have arisen concerning the relative
biological validity of tritium and iodine-125 labelled
peptides. Some workers fear the size of the iodine group
and the consequent 'isotope' effects it may have. It has
been found that although the choice of radioisotope
(tritium or iodine-125) and position of labelling may not
seriously influence receptor studies, catabolism of the
two forms may be different. Digestion of the peptide
within cells of the liver, the kidneys[10] or the blood-
stream may be influenced by the position of labelling.
Perhaps more important in degradation studies, the tracing
of peptidolysis is only possible over periods of hours
using labels placed in the molecule such that the radio
marker, preferably tritium, is retained upon the major
portion of the peptide until near the end of the
degradation process. The validity of a particular
tritiated peptide for tracing intracellular peptidolysis

is best determined by experiment. Such pilot experiments may be made either by autoradiography at the light microscope (LM) level or using radio-HPLC.[10]

It has been pointed out elsewhere[2] that although isotope effects are detectable in analytical chemical systems (e.g. various varieties of chromatography – see elsewhere in this volume), they are much harder to detect in cellular material, especially so in kinetic studies of intracellular translocation wherein the tissue structure is to be retained. The only possible way appears to be to compare the autoradiographic results obtained with two different forms of radioisotopic labelling (e.g. tritium labelled and carbon-14) of the same precursor. It must be said that in many experimental situations that it is very awkward if not impossible to do.

Doses of Radiochemicals in Biological Work

Doses vary widely depending on the purpose of the experiment and on the radioisotope used. A few examples may be a useful way of giving information. In whole body studies where perhaps the distribution of chemical is being studied within a small mammal such as the rat (say a 200g animal), as little as 5μCi may give usable results provided the isotope is carbon-14 or one of comparable energy of emission. In some studies where more time is let elapse before sacrifice, the dose would perhaps be 50μCi. Tritium on the other hand in such a situation requires much higher doses – approaching 1mCi per rat.

In cell kinetics where tritiated thymidine is greatly in use, 1μCi per gram body weight is generally recommended. Incorporation is into the genetic material and higher doses can cause chromosome breaks and indeed slow down the division cycles of the cells under study. High doses result in selective cell death by local irradiation.

Experiments at the ultrastructural level require high tissue levels, due to the very small mass of tissue present in one section. (Ultrathin sections are generally about 40-70nm thick.) Achieving such high tissue concentrations can be expensive, since one particular cell type may be studied out of a whole animal and hence much of the radiotracer will go to loci other than that to be sampled. Studies in the 200g rat, typically might require doses of 1mCi. Williams[2] has worked examples illustrating how doses for <u>in vivo</u> labelling may be calculated or exposure times deduced from LM preparations. It will not be surprising that recent years have seen a considerable growth in the use of the in vitro cell systems for radio-ultrastructural studies. Doses offered in vitro depend upon the length of time cells are to be exposed to the tracer. Long term labelling (over hours) is generally done with low doses (1-10μCi/ml of medium), whilst pulse labelling, which may be as short as 1 minute, might involve doses as high as 1mCi per ml. Efficient labelling is sometimes facilitated by the use of a tissue culture medium which is specially formulated not to include chemicals which lower the actual or effective specific radioactivity of the tracer chemical. Serum which is a common additive to media for the maintenance of animal cells may, in unfortunate circumstances, lower the specific radioactivity of tritiated amino acids by as much as 400 fold. Media for plant cell growth and maintenance are usually simpler and generally present less problems of this kind.

4 AUTORADIOGRAPH GEOMETRY

Generally autoradiographs consist of a specimen with a detector layer closely attached to one surface. Most commonly, the specimen is planar or nearly so and may consist of a section (slice) of biological matter, a ground slab (of rock, man-made material or aggregate) or a mass of material (e.g. a metal) with a polished surface. Sometimes a smear of particles (dust, spores, cells) may be used providing a geometrically less ideal object. Only occasionally are autoradiographs made in which the specimen is surrounded by detector material. However the approach would seem to have potential for the study of powders. The less than ideal geometry of smeared or suspended particulates or of freeze-fractured solids makes discussion of them awkward since the geometry is between '2π' and '4π'. Specimens with 4π geometry are best considered when image analysis and quantitation are discussed and the account given at this point is of idealised 2π geometry. For the purpose of this account β radiation range in the specimen and in the detector layer are referred to and yet no account is taken of the tortuous nature of β tracks. In addition, the spectrum of particle energies displayed by one beta emitter is necessarily overlooked. However, the account should none-theless be useful, and it is hoped that it will clarify some of the issues that arise in choosing the level to work at, choosing methods of specimen preparation, and adopting the strategy for quantitation appropriate to the questions being asked. Close contact between the specimen and the detector layer is assumed in the early part of the argument.

The flux of radiation around a small intense beta source is spherical. Depending where the source is placed in the specimen, and on the thickness of the detector layer, the flux may form either a complete or part hemisphere in the detector layer. The more complete the hemisphere in the detector, the more image information will be collected. Part of the information may be hard to interpret, since the proportion of the whole collected may be unknown. Figure 2 illustrates six geometric situations each of significantly different character. In 2a the detector layer collects 50% of the flux from the source giving the most efficient recording possible. None of the flux of radiation is lost in the specimen itself. In 2b the specimen is thicker and hence some of the radiation fails to reach the detector. The efficiency of recording is thus lower than in 2a. The precise efficiency in this sort of situation could be between 50% and a very small value. Figure 2c illustrates a situation with zero efficiency. If the recording layer over a thin source (as in Figure 2a) is made thinner then efficiency is lost when a portion of the hemisphere is cut away, see 2d and 2e. In 2e the extreme situation is illustrated wherein the detector is of minimal thickness – achieved using nuclear tracking emulsion by creating a layer one crystal thick.[2] This arrangement is the one that pertains in electron microscope specimens (see also below). Efficiency of recording which depends not only on the geometry but also on the type of recording material and its processing and on the emission energy of the electrons being recorded, may be ascertained by making autoradiographs of standard materials of known specific activity.[1,2,3] These must of course be prepared for autoradiography in a manner appropriate to the system needing calibration (section thickness, emulsion type, emulsion thickness, exposure time, developer, development etc). Efficiency is defined in the Appendix.

In order to complete the discussion of autoradiograph geometry, more figures are necessary. 2f depicts an important arrangement in which both specimen and emulsion layer are 'infinitely thick'. This means that the results are insensitive (within certain limits) to variations in specimen (section) thickness and emulsion thickness. Furthermore, different specimens prepared in this style yield direct comparisons of radioisotope concentration. In work at the light microscope level with low energy emitters such as tritium, preparations of this type are

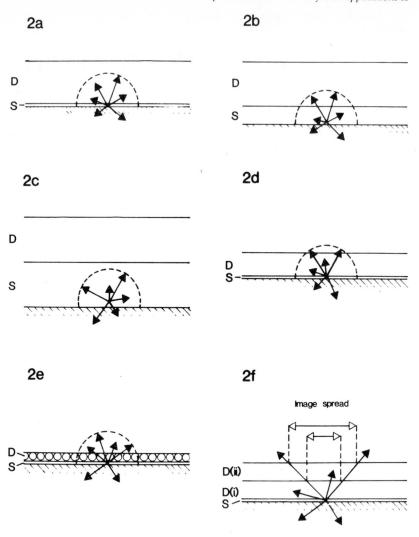

Figure 2   A series of idealised vertical sections of
autoradiographs showing various situations.   2a maximal
efficiency of recording; 2b decreased efficiency due to
self absorption; 2c zero efficiency; 2d efficiency loss
due to failure to contain the flux; 2e an electron
microscope autoradiograph with a monocrystalline emulsion
layer; 2f the effect of thickness on image spread.
Symbols: D detector layer,S specimen e.g. tissue or
material section.

invaluable and hence widely used. Using tritium, sections of 5μm thickness and emulsion of 3μm or greater are appropriate. Achieving the same situation for specimens marked with carbon-14 or sulphur-35 is almost impossible, since the beta particles can penetrate up to about 100μm in material of density of about unity.[8] Experiments employing such emitters are thus very sensitive to variations of section thickness and of emulsion thickness, both of these parameters having to be standardized most carefully.

Section Thickness

Section thickness control is an age old problem in microscopical laboratories. Modern resin embedding media such as Araldites, Epon, Vestopal, nanoplast or JB4 can be cut on high performance microtomes to much better tolerances than could be achieved with older-style wax-infiltrated specimens, or indeed can be achieved with frozen materials (e.g. human, animal or plant tissue quenched in liquid isopentane). Wax section replicates often vary in thickness by 20-40%. Section thickness can be ascertained via a variety of methods including interferometry, microinterferometry, fold diameter measurement[16] (for ultrathin sections), or more heroically by resectioning sections.[14] Surfometers have also been used.[15] The most advantageous situation is the cutting of semithin (i.e. 0.25-2.0μm) sections on an ultramicrotome. In this case the high machine tolerance intended for ultrathin cutting gives these thicker sections with a reproducibility better than 10%. Ultrathin sections on the other hand can vary greatly in thickness.

Nuclear Emulsion Thickness

Nuclear emulsion layers may also require thickness measurement. Interferometry is usable here also, but so are estimations of silver by, for instance, atomic absorption spectroscopy. Direct inspection of samples in the transmission electron microscope is the best test of thickness and crystal packing quality when working with the very thin recording layers used for electron microscope autoradiography. Stripping films (e.g. Kodak AR 10) and sheet films (e.g. CEA Verken Ultrofilm, Amersham Hyperfilm 3H) tend to be of well controlled thickness.

Role of thickness in image spread. The thickness of specimens and of recording layers have a major role in determining the lateral spread of the image. It will be apparent from the simplified drawings in Figure 2 that the

thinner the specimen and the thinner the emulsion layer
the less the image will spread or to use common
terminology: the higher the resolution. (In rigorous
quantitative studies employing refined image analysis
methods – now common in microautoradiography, the
resolution values usually quoted describe the image spread
around a straight radioactive line. The HD value[17] is the
distance from such a line that accounts for 50% of the
grains. See also Appendix.) It will easily be clear that
steps taken to increase resolution may not only decrease
the signal strength, but may also take the experiment into
a style wherein the section thickness requires careful
determination. Thus going for high resolution may demand
payment in the form of longer recording times and a
technically more awkward and capricious preparation
method.

5 TYPES OF SPECIMEN

For practical purposes it is necessary to divide specimens
(here a specimen means an object to be subjected to autoradiography) into two categories.[1,5,18] These are: (a)
one in which the labelled material is immobile and hence
not subject to extraction or translocation under the
influence of solvents; and (b) one in which the labelled
materials are indeed thus labile. A possible third
category of specimen would have volatile radiolabelled
substances present. (There was once an anaesthetist who
wished to localize diethyl ether in brain tissue!)

Categories of Specimens

     Category (a) specimens include biological ones in
which the offering of soluble precursors gives rise to
radiolabelled macromolecules which are either insoluble of
themselves or more usually can be made so by fixation
treatment with one of a range of agents. Cross-linking
aldehydes such as glutaraldehyde are the agents most
commonly used at the present time. In these specimens,
the extraction of unmetabolized soluble precursors and of
its conversion or catabolic products is useful since it
leads to an image entirely devoted to labelled macromolecular products. Other category (a) specimens are
synthetic polymers and low solubility mineralogical
specimens. Category (b) specimens would include soluble
chemicals on chromatograms, and cells and tissues
containing radioactive ions such as sodium-22, chloride-36
or calcium-45, tritiated sugars, amino acids, nucleotides
etc, and any of a great variety of radiolabeled drugs and
xenobiotics.[7] Some substances do not fit clearly into

either category. Thus radiolabelled agents used to detect
surface receptors on cells may or may not be tightly bound
to those receptors. Their extractability status is only
ascertainable by experiment using counting equipment.
Likewise, specimens of an inorganic nature may have to be
evaluated by counting. Ion exchange resin beads
containing radioactive ions have been sectioned and
autoradiographed - a procedure which is possible without
too much problem when the ions concerned have been
precipitated as insoluble salts within the beads by a
suitable treatment. Using such an approach, the gradient
and depth of the exchange processes within the beads
whilst they are in filter beds can be successfully
studied.[19]

Results on category (a) materials are reported in
thousands of articles in journals covering cell and tissue
biology, cell kinetics, dental science, neurobiology,
pathology, cancer research, plant biology and pathology
and numerous other areas. Category (b) materials are to
be found in journals devoted to pharmacology, toxicology,
endocrinology, pesticide research, agrochemicals and
others. Applications in applied chemistry, geology and
materials science[20] are valuable but occasional and
usually come under category (a) and of course are
scattered through many publications. In biochemistry, the
nucleic acid field is now the major area of application.
Generally the work is of a macro nature and it matters
little if the specimens are considered to be of either
category. In situ nucleic acid hybridization[21,22] is a
rapidly expanding field however and is in category (a).

Soluble Compound Specimens

Methods for handling category (b) specimens are much
more difficult to master than the other methods, since
specimens often require freezing in order to immobilize
the chemical contents. Freezing can do a great deal of
structural damage if it is not well handled. Fracturing
or sectioning frozen material to produce quality sections
without subjecting them to transient thawing requires
skill. Applying nuclear emulsion layers to such sections
without thawing them is also a skilled matter. For
details of these methods see[1,4]. See also articles
discussing fitting of appropriate methods to experimental
problems.[23,24] Freeze-fracture autoradiography attempts
to relate radioactivity to the topography of the fractured
specimen surface,[25] a matter not only technically
delicate, but often of doubtful value since the radiolabel
could often lie in features which are below the fracture

relief and hence not of the morphological image.

6 RECORDING MATERIALS

Nuclear tracking emulsions (table 2) are 'silver halide-in-gelatin' emulsions especially designed for detecting, tracking and quantitating emissions from radioisotopes. They are sensitive to β and α radiation and to some extent to γ radiation. They are flexible and moldable and hence very adaptable to sources of different topography. They are designed to exhibit a narrow range of crystal sizes - over 95% of the crystals lying within a 2.5-3.0 fold range. This is a much narrower range than for instance, the crystals of some panchromatic roll films. The mean crystal size for emulsion sold as gel pellicles ranges from $0.06\mu m$ in the cuboidal crystals of Eastman Kodak 129.01 to $0.3\mu m$ for the Ilford G series emulsions. Sheet films that have been sold for medical X-ray work sometimes contain large crystals measurable in microns or tens of microns. A list of some of the available products for microautoradiography is given in table 2.

Various sensitivity ratings may be available for each crystal size, the ratings (in the Ilford series ranging from 0 to 5 describing increasing levels of sensitization via additives). The sensitivity rating and the mean crystal diameter both play a role in determining the ease with which incident particles are detected. It will be obvious that larger crystals on average collect more energy from incident particles than small ones and hence are more sensitive.

Nuclear tracking emulsions are expensive (roughly £80 to £200 for a 50ml or 4oz bottle). Shelf life varies, but may be as short as 2 months, although unopened bottles maintained at 4°C are sometimes usable after 6-9 months storage. Sheet film and stripping films collect background easily and have limited life. They may easily suffer from a damp atmosphere or from environments containing reducing or oxidizing vapours. Wood sometimes gives off vapour which fogs nuclear emulsions.

Alpha particles and fission fragments both give rise to tracks in low mass materials. These fine pathways can be made visible to the light microscope by etching methods. Thin layers of various plastics or celluloses are suitable.

Application of Nuclear Tracking Emulsions to Specimens

Nuclear emulsions will melt if they are warmed above
43°C. They yield a thick creamy material that can be
diluted as desired with distilled water. When they are
cooled they set to a gel. If that is placed in a dry
atmosphere, it dries to a firm thin layer which, unlike
the water-rich gel, is capable of recording radiation.
The plasticity of the emulsion can be adjusted by the
addition of glycerol as plasticiser. Emulsions are
sensitive to physical stress such as compression and
stretching. (They thus easily record finger prints.) The
addition of a plasticiser reduces the formation of latent
images by physical insult - an important result when
working with rough surfaced specimens. Many sections of
biological material are sufficiently rough to warrant the
use of a plasticiser. Ilford emulsions differ quite
markedly from those of Eastman Kodak; in colour,
viscosity, smell and susceptibility to chemographic
effects. The Ilford materials contain significant traces
of detergent which not only smells characteristic, but
also imparts special properties. They are thus castable
on a large wire loop. In addition they foam if stirred at
all vigorously, meaning that they must be treated with
care. Emulsions of other manufacture can be cast as thin
layers provided a detergent is added.

The thin layers cast on wire loops are invaluable in
microautoradiography.[2] They can be made to contain a
closely packed layer either monocrystalline or slightly
overlapping. These can be applied with advantage to
ultrathin sections for electron microscopy or to semi-thin
sections (0.25-1.0μm) for the light microscope. Specimens
1.0-10.0μm in thickness can be coated with emulsion by
dipping them in a jar of diluted emulsion, a method in use
many years.

Layers of emulsion produced by the dipping method
mentioned above are only of even thickness when they are
generated upon highly planar specimens. This is of little
importance provided the radioisotope being localized is of
low enough emission energy such that the incident
particles are all contained within the recording layer
(tritium, iodine-125, iron-55). In the case of higher
energy isotopes however (carbon-14, sulphur-35, iron-59,
phosphorus-32) local variations in recording efficiency
will result from local changes in emulsion thickness.
Ideally, rough specimens containing these isotopes should
be studied using a sheet emulsion or stripping film layer.
Details of the methods for producing layers by dipping,

wire loops or with stripping film will be found in the
books by Rogers[1] and Williams.[2]

    Exposure. The exposure time (i.e. the time taken to
collect the image) is usually in the area of days or
weeks. Only occasionally are exposure times as short as
hours used. For this to be possible extremely high
activities are necessary e.g. >0.5mCi per gram of
specimen. When ultrathin sections are used the specimen
mass is so small that quite high activities must be
employed to gain sufficient data over exposure times of
weeks (see[2,3]).All exposures naturally take place in
light-proof boxes (black plastic is ideal e.g. Clay
Adams). Cooling to 4°C is recommended and most workers
include in each box a dust-tight capsule of drying agent
(e.g. Dry Air Corporation, U.S.A.). Dry atmosphere
discourages or prevents fading of the latent images,
whilst low temperature reduces chemography (see Appendix).

Development and Darkroom Conditions

    Development like emulsion application takes place
under darkroom conditions. For the best and most
convenient work a dedicated darkroom is required. This
should have an effective light trap to allow operator
egress when emulsions are uncovered, air conditioning, a
variety of relevant lights (and especially a sodium lamp
(e.g. Thomas Inc.), a fridge for storing coated specimens
in the room, dedicated glassware (see Figure 2), water
baths, a sink, maybe a cryostat for frozen sectioning of
tissue, supplies of masking tape and other sundries and a
portable wireless. The safe lights should have 15 watt
bulbs and be fitted 4ft above the work surfaces. Good
organization is essential.

    The developers used in microautoradiography are
almost exclusively of the lytic 'chemical' type. Although
occasional references to physical developers occur in the
literature (see [2]), no true physical developers are in
use. For discussion of the chemistry of photography see
Mees and James,[49] or Baines.[50] D19 developer, probably
the most commonly used commercial development formulation
is based on Elon which like Genol and Metol is a trade
name for monomethyl para aminophenol sulphate. This
formulation contains also hydroquinone, Na_2SO_3, Na_2CO_3 to
give a pH of 9.9 and KBr as a restrainer. The full
formulation is given in Williams.[2] Microdol X, another
commercial formulation of lower pH, is also employed in
some work, especially those experiments where less
vigorous development is required or less complex silver

grain forms needed. Recording efficiencies are lower with Microdol than with D19.[2] Kodak Dektol is widely used in the USA. Laboratory-made developers find occasional use e.g. Elon-ascorbic acid.[2] Generally such developers are of pH values near neutrality and of relatively low efficacy. The grain forms shown in Figure 1 are those given by D19 development of Ilford L4 emulsion. Other grain forms are illustrated in Williams.[2] In micro-autoradiography development time and temperature must be tightly controlled and adjusted to the requirements of the method to be used to assess the finished product. For example, in specimens to be evaluated by incident illumination, development must be restricted in order to reduce the number of minute background silver grains.[1] For macroscopic specimens standard development as specified by the film manufacturer will be quite adequate since background levels and grain size are not crucial. Fixers in general use in macroautoradiography are of course also standard. In micro work, however, hardeners are generally omitted.[2]

The Use of Scintillators ('Fluorography')

The use of fluors for enhancing autoradiographs was suggested some years ago,[26] and it has been widely employed in macroautoradiography especially of electro-phoresis gels where great shortening of exposure times has been reported to result[29] and for which purpose commercial enhancers are available e.g. AmplifyTM, Amersham International plc. Rogers[1] attributes the improvement to a reduction in latent image fading via the scintillator solvent rather than to photon amplification effects. Preflashing of sheet film has also been employed as an enhancing procedure.[1] In microfluorographic work only the method of Buchel et al.[27] appears sound and that has not been fully evaluated and exploited.

7 EVALUATION AND INTERPRETATION OF AUTORADIOGRAPHIC IMAGES

Recording

The first preliminary evaluations of an autoradio-graph will either be by eye or in the LM. However, for most analytical purposes the image is recorded as a light micrograph or electron micrograph and it is this secondary image that is used for analytical examination. Electron micrographs must be prepared at magnifications calibrated via a grating replica.[3] Appropriate micrometers must be used when recording with the light microscope. It should

be noted that LM autoradiographs are generally somewhat
thicker than the depth of field of a ×100 oil immersion
lens. Thus getting the specimen and the silver grains in
focus at the same time may be impossible. The thinner the
emulsion layer the less the problem. Modern video
recording methods if available do permit the making and
combining of two images to solve this problem.

Quantitating the Response

     Silver grains en masse confer light stopping power to
the emulsion which may thus appear grey or black. The
measurement of 'blackness', or more properly put the
optical density (the negative log of the transmittance),
provides a direct measure of exposure to incident
particles. This is an invaluable method of quantitation
for macro specimens or for rather uniform micro specimens,
and numerous pieces of densitometric equipment are
available for making the measurements (see[1,28]).

     In autoradiographs viewed under the microscope the
image breaks into discrete grains which if sparse are
difficult to quantitate densitometrically[28]. They have to
be either counted as individuals or measured as the amount
of silver metal. In most autoradiography the reflectance
of incident light stands for the amount of silver metal,
although it is in essence an estimate of coverage with
metal.[28] Many laboratories now own light microscopes
equipped with incident dark field optics and appropriate
diaphragm systems that will permit reflectance images to
be made of individual biological cells or other small
objects. Photometer attachments allow the reflectance
signal to be measured and recorded, a significant signal
that may emanate from as few as five silver grains. Since
the latter may have been collected over several weeks of
'exposure' at perhaps an efficiency of 10 or 20%, it can
be seen that activities as low as perhaps 2×10^{-4}pCi can
be reliably detected. (cf. The ordinary laboratory liquid
scintillation counter where in normal usage the limit is
$10^{-4}\mu$Ci or possiby an order of magnitude better.)

     Individual grains can be counted by eye using a ×100
oil immersion lens, a painstaking method that has the
advantage that it can be combined with many selective
biological stains. However modern image analysers can
also count grains and grain clusters, and with refined
window settings generate large volumes of grain count data
for small objects with far less strain. Goldstein and
Williams[28] and others[1,3] have discussed the relative
merits of these methods elsewhere.

It will be evident that the interpretation of quantitative image data is only possible if the following matters are taken into account:

i   The signal strength relative to the background level;
ii   The presence or otherwise of chemographic effects;
iii The selection of appropriate volumes of emulsion in which to count grains i.e. allowing for image spread.

Interplay of the image spread from adjacent sources must also be taken into account by appropriate image analysis methods. These are discussed in a later section.

Background

Background grains derive from several origins. The first of these is spontaneous latent image generation in unexposed crystals ('fogging'). This can be very important in small crystals and may curtail the shelf life of emulsions with very small crystals. Related to this, is the possibility of developing unexposed crystals by over-vigorous development (usually by using too high a temperature). Many developers have restrainers added to discourage these processes. Physical pressure may contribute to the background level too.

To be added to these sources of background are light (of course!), gamma radiation in laboratories, cosmic rays, microwave radiation from ovens and natural radio-activity in the environment. The latter includes radium-226, radon-222 and thorium-232 from building materials and, more importantly, potassium-40 in construction materials and microscope slides. Control experiments for chemography and regular background estimates are essential for critical work.

8 IMAGE INTERPRETATION

How Many Grains are Necessary to Prove One Object to be Radiolabelled?

In many fields a question repeatedly asked is 'How many grains must lie over a particular object (e.g. cell) before I can reasonably call it labelled?' The background level is the key to answering this question. The frequency of background grains must be determined e.g. by delineation of quadrats of appropriate emulsion area, counting grains in each, determining the mean grain

density and the type of distribution it follows. This
might be supposed to be Poisson, but frequently it is
better described by the negative binomial.[30] From the
mean grain density and distribution of the proportions of
cells with 0, 1, 2, 3, 4, 5, etc grains attributable to
background can be estimated. It is thence possible to set
grain count minima for presumptive radiolabelling. It is
also possible to compute background grain distribution
curves and subtract these from the frequency curves
describing experimental populations to obtain estimated
grain frequency curves for labelled populations.[31] In
skillfully prepared light microscope autoradiographs 5 or
sometimes 4 grains are sufficient to confirm labelling of
particular cells.

Defining Volumes of Emulsion to Count for Obtaining Grain Count Data which allow for Image Spread

Image spread is usually estimated for a given system
by preparing autoradiographs of straight radioactive lines
('hot lines' or 'line sources'). Since the functions that
describe image spread from lines (lspf) can be used to
yield further functions for radioactive points (point
source probability functions, pspf) it will follow that
the image spread for continuous sources of simple shape
can be computed directly. Thus image spread functions for
discs, annuli, squares and rectangles can all be made
available. Their spread may be conveniently normalized in
HD units[17,32]. Downs and Williams[33] have published tables
covering sources of various shapes and sizes and indeed a
method of 'stacking' the data from sets of sources of
similar shape but dissimilar size. Those data vividly
illustrate the manner in which small sources suffer worse
losses from the effects of image spread than do large
ones. An application of such data is given in[34] and the
effects of some departures from the presumed source shape
in[35].

Absolute Quantitation and the Use of Calibration Standards

Efficiency in autoradiography can be estimated by
comparing grain or track numbers with disintegration
numbers. Grain yield estimates can be used to relate
track numbers to grain numbers. (See Appendix for a brief
account of grain yield.) The numbers of disintegrations
in a biological or 'materials' specimen can be ascertained
destructively after autoradiography by digestion and
scintillation counting or where appropriate by use of a
well counter.

Reference standards of established specific activity can be used to ascertain the activity of experimental specimens by comparison i.e. sections of standard specimens are autoradiographed under identical conditions to experimental sections and the results compared. Commercially available multi-activity micro standards of appropriate dimensions designed by the author are marketed by Amersham International plc.[36]

4π Specimens

In some fields it is advantageous to embed the specimens in the detector material. Levinthal and Thomas[53] showed that it was a valuable approach in microbiology. The study of sediments and powders would also be facilitated. Track counting of such preparations yields excellent quantitative evaluation.

Analysing and Interpreting Autoradiographs in which Images Spread over Adjacent Features

Significant lateral spread of the image results in grains appearing over features other than those which emitted the particles in question. This phenomenon is known as cross-fire. It will be clear after some perusal of the account given of sources of simple shape (see above and reference [33]), that sources will vary in image loss due to spread depending on their diameter. Mixtures of large and small sources will therefore be very hard to interpret. However, the matter is even more complex, since the shape of sources profoundly affects the degree of spread and the packing of sources affects the numbers of grains acquired by cross-fire. In complex specimens assumptions about source shape are only seldom useful. These images must be analysed by a procedure that at its root breaks each putative source into an array of tiny pieces (in practice 'intersections' from superimposed line pattern, 'points') which are in some senses equivalent to the pixels of an image analyser screen. The potential cross-fire from these points is assessed by sampling using an overlay screen,[37,38] or if in an image analyser by line array generation.[39] The superimposed points have attached to them random directions (chosen from 16 possible) and distances drawn in an unbiased fashion from the pspf scaled to the appropriate magnification for the experiment. The resultant notional cross-fire data are collected as a matrix.[37] The real silver grain data can then be corrected for cross-fire using the matrix by solving a set of simultaneous equations,[37] by a Chi[2] minimizing routine or by regression methods.[40,41] Downs and Williams[33] prescribe the use of square matrices to describe cross-fire, Blackett and Parry[40] rectangular ones. When tested on replicate experimental material these approaches give closely similar results. A comparison of the cross-fire matrix method and analyses based assumptions of source shape for one cell biological system are given in [34].

A method which grows out of the above is reported by Miller et al.[42] It is a maximum likelihood method constructed for use in image analysers. It is in the longer term a likely way forward but it does require very high quality hot line data. These are at present in limited supply. Particularly needed are data on the reproducibility of hot line experiments. This will soon be available for semi-thin sections containing hot lines of iodine-125.[43] Few other relevant data are on the horizon. The great majority of published hot line data refer to single experiments or at least unstated numbers of preparations.

The role of anatomical models in image interpretation. Heavy autoradiographic images i.e. those which are macroscopically grey or black, can be interpreted in the absence of structural information about the specimen. However sparser images cannot be interpreted without the creation of some model of the anatomy of the specimen. Thus in a specimen yielding sparse images, to ask if the interior of a cell is uniformly labelled requires the division of the cell into at least two items. These are usually chosen on some biofunctional basis e.g. nucleus and cytoplasm, although this need not be the case. As pointed out elsewhere,[44] the limits in the extraction of information from complex specimens lie in (a) the signal strength, (b) the resolution and (c) the ability to create anatomical models. The last of these is often the most crucial factor though frequently it is not appreciated to be so. Experimenters often discuss the importance of the resolution of their autoradiographs and fail to appreciate that whilst resolution is important, anatomical delineation is crucial. The discrimination possible in an autoradiograph depends on the anatomical delineation which in turn prescribes the available modelling that is the base of refined modern analytical methods. Modern methods can discriminate label in features smaller than the HD value of the system.[51,52]

Model testing. It is evident that the local specific activity estimates obtained from a complex specimen are data that are attributable to the particular model used to make the analysis. These data are only of value if the model is a valid one. Models must therefore be tested. Downs and Williams[37] have pointed out that the complete analytical process must consist of model creation, local activity estimation based on the model, model testing, model adjustments as deemed appropriate, local activity estimation based on the new model, model testing again. Thus the best model must be approached via an iterative process.

The method of Blackett and Parry,[40] which uses rectangular matrices to collect cross-fire data, permits some remodelling of junctions (interfaces between components) without major remodelling. This approach and that of Downs and Williams[37,38] give extremely similar answers though each in their own way.

9 THE FUTURE

There has been a steady growth in the use of autoradio-graphic methods over the last 25 years. Currently receptor studies and in situ hybridization are especially prominent. The future will see further developments of these and other areas – a process stimulated and supported by the use of image analysing computers.

REFERENCES

1. A.W. Rogers, 'Techniques of Autoradiography', Elsevier, Amsterdam, 1979, 3rd edition.
2. M.A. Williams, 'Autoradiography and Immuno-cytochemistry, Practical Methods in Electron Microscopy 6, Part I', ed. A.M. Glauert, North Holland Press, Amsterdam, 1977.
3. M.A. Williams, 'Quantitative Methods in Biology, Practical Methods in Electron Microscopy 6, Part II'. ed. A.M. Glauert, North Holland Press, Amsterdam, 1977.
4. R. Baserga and D. Malamud, 'Autoradiography, Techniques and Application', Harper and Row, New York, Evanston and London, 1969.
5. L.J. Roth and W.E. Stumpf, 'Autoradiography of diffusible substances', Academic Press, New York and London, 1969.
6. D. Green, G. Howells and M.C. Thorne, <u>Phys. Med. Biol.</u>, 1977, <u>22</u>, 284.
7. Proc. Intern. Conf. on Autoradiography, Sheffield, <u>Proc. R. microsc. Soc.</u>, 1986, <u>21</u> No. 5.

8.   M.A. Williams, J. Microsc. (Oxford), 1987, 145, RP1.
9.   E.A. Evans, 'Tritium and its compounds', Butterworth, Oxford, 1974.
10. J.I. Lowrie, J.R.J. Baker and M.A. Williams, 'Renal heterogeneity and target cell toxicity', ed. P.H. Bach and E.A. Lock, Wiley, New York and Chichester, 1985, p. 63.
11. E.J. Hahn, 'Photographic procedures for light and electron microscope autoradiography. A selected bibliography of books and papers', Eastman Kodak Co., Rochester, New York, 1984.
12. M.A. Williams and G.A. Meek, J. R. microsc. Soc., 1966, 85, 337.
13. G.F.J.M. Vrensen, J. Histochem. Cytochem., 1970, 18, 278.
14. D. Frosch and C. Westphal, J. Microsc. (Oxford), 1984, 137, 177.
15. A.D. Pearse and R. Marks, J. Clin. Path., 1974, 27, 615.
16. M.A. Williams, Stereol. Iugsl., 1981, 3, suppl. 1, 369.
17. M.M. Salpeter, L. Bachmann and L.L. Salpeter, J. Cell Biol., 1969, 41, 1.
18. M.A. Williams, Adv. Opt. Electr. Microsc., 1969, 3, 219.
19. D. Petruzelli, personal communication.
20. R.H. Condit, 'Autoradiographic techniques in metallurgical research', in 'Techniques of Metals Research 2, Part 2', ed. R.F. Bunshan, Wiley, New York.
21. M. Jamrich, K.A. Mahon, E.R. Garvis and J.G. Gall, EMBO J., 1984, 3, 1939.
22. D. Myerson, R.C. Hackman, J.A. Nelson. D.C. Ward and J.K. MacDougall, Human Path., 1984, 15, 430.
23. M.A. Williams, 'Nephrotoxicity in the experimental and clinical situation, Part 1', Martinus Nijhoff, Boston and Dordrecht, 1987, p. 1.
24. J.R.J. Baker, 'Nephrotoxicity in the experimental and clinical situation, Part 1', Martinus Nijhoff, Boston and Dordrecht, 1987, p. 85.
25. T.D. Le and B.E. Wilde, Corrosion-NACE, 1983, 39, 258.
26. B.G.M. Durie and S.E. Salmon, Science, 1975, 190, 1093.
27. L.A. Buchel, E. Delain and Bouteille, M. J. Microsc. (Oxford), 1978, 112, 223.
28. D.J. Goldstein and M.A. Williams, J. Microsc. (Oxford), 1971, 94, 215.
29. B.M. Kopriwa, Histochemistry, 1980, 68, 265.
30. K.S. Bedi and D.J. Goldstein, Histochemistry, 1978, 55, 63.

31. H. Korr and B. Schultze, <u>Appl. Histochem.</u>, 1985, <u>29</u>, 165.

32. M.M. Salpeter and L.L. Salpeter, <u>J. Cell Biol.</u>, 1971, <u>50</u>, 324.

33. A.M. Downs and M.A. Williams, <u>J. Microsc. (Oxford)</u>, 1984, <u>136</u>, 1.

34. M.A. Williams, A.M. Downs and E. Junger, <u>Acta Stereol.</u>, 1985, <u>4/2</u>, 115.

35. E.A. Marshall and M.A. Williams, <u>Acta Stereol.</u>, 1985, <u>5/2</u>, 207.

36. Anon., 'Autoradiographic I-125 microscales BAS H/11753', Amersham International plc, Amersham, UK, 1986.

37. A.M. Downs and M.A. Williams, <u>J. Microsc. (Oxford)</u>, 1978, <u>114</u>, 143.

38. M.A. Williams and A.M. Downs, <u>J. Microsc. (Oxford)</u>, 1978, <u>114</u>, 157.

39. G. Hejblum, A.M. Downs and J.P. Rigaut, <u>Acta Stereol.</u>, 1985, <u>5/2</u>, 245.

40. N.M. Blackett and D.M. Parry, <u>J. Cell Biol.</u>, 1973, <u>57</u>, 9.

41. D.V. Markov, <u>J. Microsc. (Oxford)</u>, 1986, <u>144</u>, 83.

42. M.I. Miller, K.B. Larson, J.E. Saffitz, D.L. Snyder and L.J. Thomas, <u>J. Electr. Microsc. Techn.</u>, 1985, <u>2</u>, 611.

43. C.W. Anderson, E. Junger and M.A. Williams, <u>J. Microsc.</u>, 1988, in press.

44. M.A. Williams, <u>J. Microsc. (Oxford)</u>, 1982, <u>128</u>, 79.

45. W.A. Aherne, R.S. Camplejohn and N.A. Wright, 'An introduction to cell population kinetics', Arnold, London, 1977.

46. C. Vanroelen, F. Harrisson, L. Andres and L. Vakaet, <u>J. Microsc. (Oxford)</u>, 1981, <u>122</u>, 1.

47. S. Ullberg, 'The technique of whole body autoradiography', <u>Science Tools</u>, special issue, 1977.

48. C.G. Curtis, S.A.M. Cross, R.J. McCulloch and G.H. Powell, 'Whole Body Autoradiography', Academic Press, London, 1981.

49. C.E.K. Mees and T.H. James, 'The theory of the photographic process', Macmillan, New York, 1966, 3rd edition.

50. H. Baines, 'Photography for the scientist', Academic Press, London and New York, 1968.

51. C. Kent and M.A. Williams, <u>J. Cell Biol.</u>, 1974, <u>60</u>, 554.

52. H-D. Dellmann, J. Boudier, F. Courand, P. Cau and J-L. Boudier, <u>Neurosci. Letters</u>, 1983, <u>35</u>, 71.

53. C. Levinthal and C.A. Thomas, <u>Biochim. Biophys. Acta</u>, 1957, <u>23</u>, 453.

APPENDIX - Terms used in Autoradiography

Anatomical model: The features into which a specimen is divided for image analytical purposes.

Background: Silver grains or tracks which appear on the autoradiograph but which do not originate from the radioactive subject of the experiment.

Cell kinetics: Numerical study of cell populations including their generation, replacement and differentiation.[45]

Chemography: Production or loss of an image in an emulsion layer due to the action of chemicals in the specimen or occasionally in the environment in the vicinity of the specimen. Negative chemography is the loss of images. Positive chemography is the creation of images.

Cross-fire: The appearance of silver grains overlying a feature within which the responsible decays did not occur. It results from image spread.

Cross-fire matrix: A matrix of values describing the probabilities of cross-fire between all the various components of an anatomical model.

Densitometry: Measurement of the response of an emulsion layer as blackening. The negative log of the transmittance of light. It is proportional to electron exposure.

Development: Chemical processing in order to convert latent images into much larger silver grains.

Dipping method: Procedure for preparing autoradiographs by dipping specimens in liquid emulsion.

Emulsion: A gelatin-silver halide emulsion designed to record ionizing radiation.

Exposure: The process of collecting the latent images. Usually measured in days, weeks or months.

Exposure time: The length of exposure.

Efficiency: The proportion of decays occurring within the specimen that are recorded in the recording layer. They may be recorded in nuclear emulsions as either single silver grains or as tracks. The latter may be from 2 to >70 grains long.

Fading: Loss of latent images. Promoted by water, oxidising agents.

Fluorography: Autoradiography involving the additional use of scintillation fluid or plastic.

Fog: Spontaneous generation of developable images in emulsion in the absence of light, radioactivity or chemographic effects.

Grain yield: The average number of silver grains per electron strike on the recording layer.

HR value: The distance within which half of the silver grains fall from a point radioactive source. This parameter which is only estimable indirectly equals HD × 1.73.

Image analyser: Image analysing computer e.g. Quantimet 970, Ibas II, ...

Image spread: The lateral spread of the autoradiographic image due largely to geometric considerations.[45]

Infinite thickness: Specimen thickness beyond which no further image accrues.

Loop method: Manufacture of semi-gelled monocrystalline emulsion layers on a large wire loop usually made of Nichrome.

Microautoradiography: Autoradiographs made for viewing with a microscope.

Morphometry: The measurement of structural form.

Reflectance photometry: Measurement of the amount of incident light reflected back up the tube of the microscope.

Resolution: A measure of lateral image spread for a particular geometry of autoradiograph. The measure usually employed is the distance from a straight line within which half of the grains fall. This last measure is also called the HD value.

Stereology: A collection of methods, applicable in many fields, for collecting two-dimensional information and computing from it three-dimensional information about the same object or similar object.

Silver grain: A complex structure composed of silver metal. (See Figure 1.) It appears as a dense black dot when viewed in bright field illumination. One grain or track of grains records a single electron strike.

Stripping emulsion: An emulsion prepared as a tough layer on a glass plate and intended to be peeled off for use.

Track: A row of silver grains indicating the path through the emulsion of one β or α particle.

Visual grain counting: Evaluation of an image by literal counting of the silver grains by human eye and hand.

Whole body autoradiography: Autoradiography of slices of whole animals - much used in the pharmaceutical industry.[47,48]

6
Isotope Shifts in NMR Spectroscopy — Measurement and Applications

D.B. Davies, J.C. Christofides, and R.E. Hoffman

DEPARTMENT OF CHEMISTRY, BIRKBECK COLLEGE, MALET STREET, LONDON, WC1E 7HX, UK

1 INTRODUCTION

Isotope effects are well established in NMR spectroscopy[1,2] and are used for a wide range of studies, e.g. structural analysis, details of reaction mechanisms and metabolic pathways, testing theories of chemical shifts. The use of NMR spectroscopy in the structural analysis of labelled compounds has been covered by F.M. Kaspersen in the present volume and my discussion will concentrate on NMR experiments involving replacement of hydrogen by deuterium, especially for molecules with readily exchangeable hydrogen atoms. The NMR experiments are applicable, of course, to isotopic replacement of other magnetic nuclei.

Replacement of hydrogen by deuterium is a very common procedure in NMR spectroscopy causing three changes in the proton spectrum; the disappearance of signals, the effective disappearance of appropriate spin coupling constants and a small chemical shift change on observation of the primary or, more usually, the secondary nucleus i.e. an isotope effect on nuclear shielding.

Primary isotope shifts are defined as the difference in magnetic shielding of the isotopes of nuclei of the same element, i.e. $\delta(D)-\delta(H)$, and are usually small in organic compounds, <0.03 ppm. Measurement of primary isotope shifts requires NMR observation of each isotope which obviously limits the use of the effect; in addition care has to be taken in the measurement and referencing of such small effects. Large primary isotope shifts (H/D up to 0.7 ppm, H/T up to 1 ppm) have been observed in hydrogen-bonded systems and have been used to assess the shape of the hydrogen-bond potential-energy function, in

particular to distinguish between single and double-
minimum potentials.[3,4]

<u>Secondary</u> <u>isotope</u> <u>shifts</u> are the differences in magnetic
shielding of the secondary nucleus being observed (X)
resulting from isotopic substitution at another site in
the molecule;

$$\text{i.e. } {}^{n}\Delta X (D) = \delta (XD) - \delta(XH)$$

where X = the secondary nucleus (^{1}H, ^{13}C <u>etc</u>) and
n = the number of bonds separating the observed nucleus
and the isotope.* The definition is exemplified by
replacement of the lighter isotope (H) by the heavier
isotope (D) which generally results in upfield shifts of
the secondary nucleus <u>i.e.</u> a negative isotope effect.   A
few positive isotope effects have been observed,[2,5,6]
particularly associated with hydrogen bonding.

As a result of experiment the effects of isotope
substitution on nuclear shielding can briefly be
summarised,[1,2]
  i)   a single isotopic substitution by the heavier
isotope produces an increase in the shielding of the
observed nucleus (secondary isotope effect) and the
nucleus of the heavier isotope has a higher shielding than
the lighter nucleus it replaced (primary isotope effect).

ii) the magnitudes of secondary isotope shifts decrease
as the site of substitution becomes more remote from the
nucleus of interest.

iii) the magnitude of the shift is a function of the
resonant nucleus reflecting the range of chemical shifts
observed for the nucleus.

  iv) isotope shifts are largest for the largest fractional
changes in mass upon substitution.

  v) the magnitude of an isotope shift with respect to the
lightest isotopomer (HH, HD are isotopomers) is
approximately proportional to the number of nuclei of a
specific type which have been isotopically substituted.

Isotope shifts originate from the isotopic dependence
of the molecular geometry and of the intermolecular
interactions which affect shielding (such as hydrogen-
bonding, neighbour-molecule magnetic anisotropy, Van der

----
* The sign convention, similar to that used for substituent
effects, is opposite to that used in a recent review.[2]

Waals forces etc). For a given state of rotation and vibration the two isotopomers will lie at different levels on the potential surface so that the degree of averaging of the nuclear shielding (σ) will be different and consequently the shielding of the corresponding nuclei will be different.[7]

2. MEASUREMENTS OF H/D ISOTOPE EFFECTS BY NMR

Replacement of H by D often requires chemical synthesis which might inhibit the use of the method either because of difficulties involved in synthesis or because the starting materials are precious. For molecules with readily exchangeable hydrogen atoms (e.g. in OH, NH, NH_2 groups) replacement of H by D (or T) is usually straightforward and can lead to extremely powerful ^{13}C NMR methods of structural analysis i.e. DIS NMR (Differential Isotope Shifts)[8-10] and SIMPLE NMR (Secondary Isotope Multiplets of Partially Labelled Entities).[11-14] Although SIMPLE and DIS NMR have been used separately for ^{13}C signal assignment,[1,3] it is found that a combination of the methods enables ^{13}C NMR signals to be assigned in H_2O, D_2O and DMSO solutions from the results of two experiments - and the methods are most powerful for molecules with many exchangeable hydrogen atoms such as the carbohydrate maltitol.[15]

A most interesting application of SIMPLE NMR is the study of OH...OH intramolecular hydrogen bonding in carbohydrates using the proton as the secondary reporter nucleus.[5,6] Observation of isotope effects on OH signals provides information not only on the presence of the hydrogen bond but also an indication of its relative strength and direction. For example, application to sucrose derivatives in the present work shows that an intramolecular hydrogen bond stabilises a weaker hydrogen-bond network in the molecule and that the whole network is cooperative. i.e. the intramolecular hydrogen bond and the network are stronger as more hydrogen bonds take part in the network.[16]

DIS NMR (Differential Isotope Shifts)

Pfeffer and co-workers[8-10] measured the deuterium-induced ^{13}C differential isotope shift (DIS) of a number of mono- and disaccharides using a special dual coaxial cell with equal concentrations of carbohydrate dissolved in H_2O (inside tube) and D_2O (outside tube); the difference in chemical shifts of the compound in D_2O and H_2O is observed in the fast exchange condition. Results

from a number of compounds were analysed statistically to
determine the magnitudes of the two (β) and three (γ) bond
isotope effects which contribute to the total shift
observed and were used in the assignment of many
carbohydrates as well as to correct some assignments of a
number of carbohydrates previously measured. The method
has limited applicability because all hydroxy-bearing
carbon atoms will appear as doublets irrespective of their
environment, because magnitudes of β and γ effects have
to be made from measurements on a series of closely
related compounds and because other isotope effects, such
as those due to hydrogen-bonding, are not differentiated
by the DIS method.

The procedure for routine DIS NMR measurement is as
follows: samples are divided into two parts, one is
dissolved in H_2O and the other, freeze dried from D_2O, is
then dissolved in D_2O. The H_2O solution is placed in a
5mm NMR tube positioned centrosymmetrically, using PTFE
spacers, into the D_2O solution in a 10mm tube. The NMR
spectrum theoretically displays two peaks for each carbon,
one of the D_2O solution and the other of the H_2O solution.
The separation between the two peaks corresponds to the
sum of the isotope effects on the resonance plus the bulk
magnetisation effect of the D_2O that surrounds the H_2O.
For a 5mm tube (H_2O) in a 10mm tube (D_2O) the bulk
magnetisation effect is usually about +30ppb (in the
direction opposite to the isotope effect) and has the same
effect on all resonances. The bulk magnetisation effect
was calculated by comparison with the SIMPLE NMR spectra
of a number of disaccharides in $(CD_3)_2SO$ solution by
averaging over all signals, according to the following
formula,[14]

$$\text{Bulk magnetisation} = \frac{\text{Total DIS effects} - \text{Total SIMPLE effects}}{\text{Total number of carbons}}$$

The effects of H/D isotopic substitution on chemical
shifts (up to *ca* 0.2 ppm)[11],[14] are hardly noticed in
routine ^{13}C NMR spectroscopy and published chemical shifts
are not normally 'adjusted' to account for the isotope
effect between measurements in, say, H_2O or D_2O even
though fully H and fully D isotopomers are being observed
in the two cases; similarly, chemical shifts in H_2O/D_2O
mixtures are weighted averages of those for H and D
isotopomers observed under conditions of fast exchange.

SIMPLE NMR (Secondary Isotope Multiplets of Partially Labelled Entities)

If a molecule with readily exchangeable hydrogen atoms is partially deuteriated (OH : OD <u>ca</u> 1:1) and measured under conditions of slow exchange (e.g. in DMSO solution), all possible ^{13}C NMR "isotopomers" may be observed; isotopomer multiplets are analysed in terms of two bond (β) and three bond (γ) isotope effects and, hence, report on the number and type of exchangeable hydrogen atoms as neighbours of the secondary nucleus. The number of signals within the SIMPLE ^{13}C NMR multiplets depends on the number of possible isotopomers and their degeneracies, and the intensity distribution depends on the deuteriation ratio.[11,12] It is found that magnitudes of β and γ effects vary with molecular structure and so are useful for assignment of signals.[11,12]

Characteristic patterns of isotopic multiplets observed in SIMPLE ^{13}C NMR spectra of alcohols and carbohydrates are summarised in Figure 1 and depend on the number (0, 1, 2 and 3 effects) and relative magnitudes of β and γ isotope effects.

<u>One isotope effect</u>. The ^{13}C NMR spectrum of partially deuteriated (OH:OD) methanol under conditions of slow exchange consists of two lines ($\Delta\delta$ =-0.126 ppm, β effect) corresponding to the H and D isotopes <u>i.e.</u> there are two different kinds of molecules in solution one of which has a protio-(CH_3OH) and the other a deuterio- (CH_3OD) substituted hydroxyl. The relative intensity of the two lines is directly proportional to the deuteriation ratio since this determines the relative populations of the protio- and deuterio-isotopomers. For ethanol (CH_3CH_2OH) a two bond effect is expected for the carbon atom to which the hydroxyl is attached (β) and a three bond effect is expected for the neighbouring carbon atom (γ). Again two lines are expected for each carbon atom but the magnitudes of the effects are -0.122 ppm for the β and -0.058 ppm for the γ effect.

<u>Two isotope effects</u>. For alcohols with two hydroxyl groups <u>e.g.</u> ethylene glycol, CH_2OHCH_2OH, there are now four possible isotopomers present in solution at the same time, the (OH OH), (OH OD), (OD OH) and (OD OD) isotopomers. If the magnitudes of the isotope effects are different, then four lines are observed whereas if the magnitudes of the two effects are the same within the resolution of the experiment, only three lines are expected, and observed signal intensities (1:2:1; OH:OD

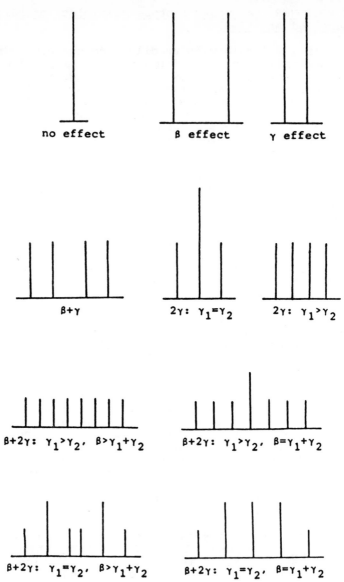

Figure 1.   Isotopic multiplet patterns observed in SIMPLE ^{13}C-NMR spectra (OH :OD, 1:1)

1:1) are accounted for by the degeneracy of the HD and DH isotopomers (Figure 1).

The relative intensity of any resonance signal is proportional to the product of the probabilities of hydroxyl groups being observed as OH or OD, i.e. p(OH) and p(OD). For signals with two isotope effects of different magnitudes the relative intensities (I) of the lines depend on the deuteriation ratio as follows:

$$I(HH) = p(OH)^2$$
$$I(HD) = I(DH) = p(OH) \times p(OD)$$
$$I(DD) = p(OD)^2$$

Three isotope effects. For carbon atoms with three isotope effects of different magnitudes, a total of eight lines are expected corresponding to isotomers with no D substitution (HHH), one D (HHD, HDH, DHH), two D (HDD, DHD, DDH) and three D atom substitution of hydroxyl groups (DDD). Depending on the magnitudes of the isotope effects it is possible to observe signals with eight, seven, six, five and, if all effects have the same magnitude a minimum of four lines. The origin of the resonance signals and their possible degeneracies are shown in Figure 1 for the three isotope effect cases often observed in carbohydrates in which the carbon atom has one relatively large effect (labelled β) and two relatively small effects (labelled γ_1 and γ_2).

For three isotope effects the relative intensities of signals depend on the deuteriation ratio in the following manner:

$$I(HHH) = p(OH)^3$$
$$I(HHD) = I(HDH) = I(DHH) = p(OH)^2 \times p(OD)$$
$$I(HDD) = I(DHD) = I(DDH) = p(OH) \times p(OD)^2$$
$$I(DDD) = p(OD)^3$$

The extent of the deuteriation can be measured directly from the proton spectrum, or from the relative intensities of lines corresponding to the H and D isotopomers for a carbon atom with only a β or a γ effect.

It should be noted that determination of the number and magnitudes of isotope effects is best accomplished at OH:OD ca 1:1 (otherwise outer lines in signals with, say, three isotope effects are very small) whereas the signs of isotope effects must be determined at other deuteriation ratios. All H/D isotope effects observed for ^{13}C NMR are negative though one positive isotope effect has been

reported for $\underline{D}$-idopyranose.[17]   In this work all
isotope effects on ^{13}C NMR spectra are negative and
magnitudes are quoted in ppb (<u>i.e.</u> 10^{-3} ppm).

3.   APPLICATION OF SIMPLE NMR IN ASSIGNMENT OF ^{13}C
     SIGNALS. <u>e.g.</u> Methyl α-$\underline{D}$-galactopyranoside.

The 100MHz ^{13}C SIMPLE NMR spectrum of partially
deuteriated methyl α-$\underline{D}$-galactopyranoside in DMSO-d_6
solution is shown in Figure 2 at different deuteriation
ratios, (a) OH:0D <u>ca</u> 1:1 and (b) OH:OD <u>ca</u> 1:3.   The
spectra consist of a series of multiplets with different
numbers of lines (1,2,3,4 and 8) and intensity ratios
which also vary characteristically with the OH:OD ratio
and the numbers of isotope effects.   The isotope effects
expected for each carbon atom of the monosaccharide are
summarised in Figure 2 using the numerical subscript
notation for the ^{13}C signal being observed (<u>e.g.</u> β_6) and,
when appropriate, a second numerical subscript is used for
the hydroxyl group which gives rise to the isotope effect
(<u>e.g.</u> γ_{32} and γ_{34} correspond to the three bond isotope
effects observed on the C3 signal from partial
deuteriation of the OH2 and OH4 groups, respectively).

    Carbon atoms with no isotope effects are observed as
single resonance signals (<u>e.g.</u> OCH$_3$ in Figure 2) and are
independent of the OH:OD ratio.   For carbon atoms with
only one isotope effect (C1 and C6) the observed doublets
are readily analysed in terms of H and D isotopomers and
confirm that β effects are usually larger than γ effects
<u>i.e.</u> β_6 = -0.118 ppm and γ_{12} = -0.016ppm.   For carbon
atoms with two possible isotope effects (C-2, C-4 and C-5)
four resonance signals are expected if the magnitudes of
the isotope effects are different (C-2 and C-4 both have β
and γ effects) whereas three resonance signals are
expected if the isotope effects have approximately equal
magnitudes (C-5, γ_{54}=γ_{56} <u>ca</u> -0.026 ppm).   Although the C-2
and C-4 resonance signals have four lines they can be
differentiated by the magnitudes of the γ effects which
depend on stereochemistry of the hydroxyl groups <u>i.e.</u>
β_2 = -0.108 and γ_{23} = -0.039 ppm whereas β_4 = -0.110 and
γ_{43}= -0.021 ppm.   The intensities of the four isotopomers
(HH, HD, DH and DD) are approximately equal for OH:OD <u>ca</u>
1:1 but exhibit relative intensities <u>ca</u> 1:3:3:9 for OH:OD
<u>ca</u> 1:3.

    The C-3 signal of methyl α-$\underline{D}$-galactopyranoside consists
of eight lines corresponding to the eight isotopomers

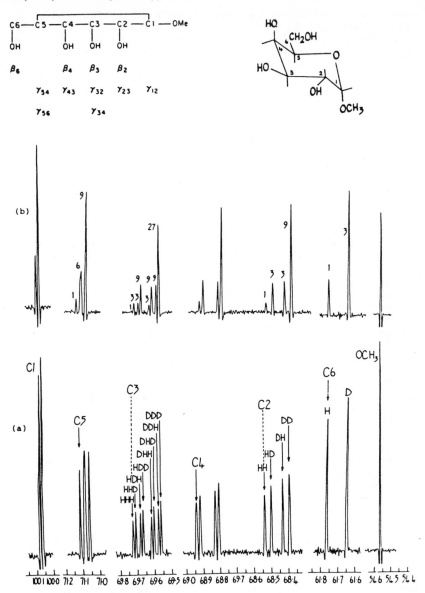

Figure 2   100-MHz proton noise-decoupled ^{13}C NMR spectrum of methyl α-<u>D</u>-galactopyranoside in DMSO-<u>d</u>$_6$ solution with partially deuteriated hydroxyl groups (a) OH:OD <u>ca</u> 1:1, (b) OH:OD <u>ca</u> 1:3

resulting from three isotope effects of different
magnitudes <u>i.e.</u> β_3 = -0.105, γ_{32} = -0.042 and γ_{34} = -0.015
ppm. The difference in magnitudes of γ isotope effects
depends on stereochemistry of the hydroxyl group with the
magnitude for equatorial hydroxyl groups (γ_{32}) larger than
for axial hydroxyl groups (γ_{34}). The intensities of the
eight isotopomers (HHH, HHD, HDH, DHH, HDD, DHD, DDH and
DDD) are approximately equal for OH:OD <u>ca</u> 1:1 but exhibit
relative intensities 1:3:3:9:3:9:9:27 for OH:OD <u>ca</u> 1:3.

   <u>Magnitudes of isotope effects</u>. Using methyl-α-D-
galactopyranoside as an example, it can be seen that the
magnitudes of 2β isotope effects are found to differ
slightly for primary (<u>ca</u> -0.118 ppm) or a secondary
hydroxyl group (<u>ca</u> -0.108 ppm) whereas a greater variation
is observed in the magnitudes of γ effects (-0.015 to
-0.042 ppm) predominantly determined by the stereochemistry
of the hydroxyl group. Similar observations have been
made for the γ_{21} effect of α and β anomers (<u>e.g.</u> γ_{21} for
α-glucose is -0.037 ppm whereas for β-glucose it is -0.067
ppm). Magnitudes of the γ_{21} effects associated with
anomeric hydroxyl groups are much larger than analogous
effects associated with other axial and equatorial
hydroxyl groups. It is the dependence of the magnitudes
of β and γ isotope effects on substitution and
stereochemistry that makes the SIMPLE NMR method a useful
aid for assignment of ^{13}C spectra of a wide variety of
molecules with exchangeable hydrogen atoms.[11-14]

   The substituent dependence of the magnitudes of the β
and γ effects has been studied using a series of simple
alcohols (Figure 3). The magnitudes of β effects of CH_2OH
groups do not depend on the substitution pattern of the
adjacent carbon atom whereas magnitudes of the γ effects
are affected by the substitution of the carbon atom
(underlined in Figure 3) for which the effect is being
observed, <u>i.e.</u> decrease of <u>ca</u> 0.009 ppm is observed for
each hydrogen substituted by a methyl group for $\underline{C}XYZCH_2OH$
type carbon atoms.

   4. COMBINATION OF SIMPLE AND DIS NMR IN ASSIGNMENT OF
      ^{13}C NMR SIGNALS <u>e.g.</u> Maltitol

Both the DIS[8-10] and SIMPLE[11-13] NMR methods have been
used to aid the assignment of ^{13}C signals in DMSO-<u>d</u>$_6$ and
D_2O solutions, respectively. There has been one report of
the combined use of the SIMPLE and DIS methods to assign
the ^{13}C NMR signals of eight glucodisaccharides in both
DMSO-<u>d</u>$_6$ and D_2O solutions[14]; it was found that nearly all

the signals could be assigned unequivocally (ca 90%) and the remaining signals are choices between two possible assignments. Recently the ^{13}C NMR spectrum of maltitol has been assigned[15] in D_2O solution using 2D-INADEQUATE NMR (Incredible Natural Abundance Double Quantum Transfer Experiment)[18,19] and in DMSO-d_6 by combination of the SIMPLE and DIS NMR methods.[15-6] Hence, it provides a good example for demonstrating the scope and limitations of the SIMPLE and DIS NMR methods for assignment of signals of compounds in DMSO-d_6 and D_2O solutions because the results can be checked with the assignment by 2D-INADEQUATE methods. It should be noted that a previous ^{13}C NMR assignment[20] of maltitol at 25 MHz had some incorrect assignments, probably due to using sorbitol as a model compound with an assignment later proved incorrect.[21]

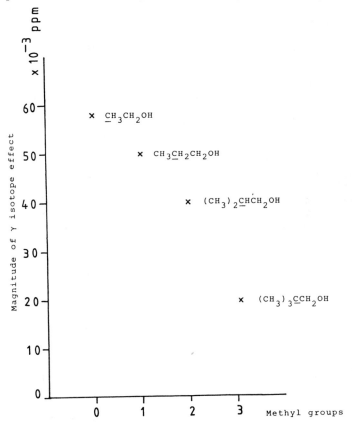

Figure 3. Substituent dependence of γ isotope effects

The structure of maltitol consists of two carbohydrate residues (Fig. 4): an α-_D_-glucopyranosyl residue (atom numbers primed) linked to sorbitol at the 4-position (atom numbers unprimed). The 100 MHz SIMPLE ^{13}C NMR spectrum of partially deuteriated maltitol in DMSO-_d$_6$_ solution (OH:OD _ca_ 1:1) is shown in Figure 4 with a greatly expanded chemical shift scale. The 12 carbon signals appear as a series of multiplets (singlet to octet) that exhibit intensity ratios depending on the OH:OD ratio. The observed multiplets are analysed in terms of the combinations of β and γ effects expected for the 12 carbon atoms of maltitol (Figure 4). In principle unequivocal assignment is possible for some signals by inspection of the spectrum because they have unique patterns of isotope effects _i.e._ C-1'(γ), C-6'(β), and other signals can be differentiated by similar patterns of isotope effects _i.e._ 2γ for C-4 and C-5', (β+2γ) for C-2 and C-3' and (β+γ) for C-1, C-3, C-5, C-6, C-2' and C-4'. A further aid in assignment is the expected values of chemical shifts (within, say, 1ppm) and magnitudes of isotope effects for the α-_D_-glucopyranosyl residue in compounds related to maltitol such as maltose and methyl-α-_D_-glucopyranoside;[11] this enables C-4 (83.9 ppm) to be differentiated from C-5' (73.3) and C-2 (72.3) to be differentiated from C-3' (73.6 ppm), especially when magnitudes of isotope effects for C-3' (90, 44, 22 ppb) and C-5' (22, 22 ppb) are compared with those for C-2 (88, 17,17) and C-4 (35, very small), respectively. Difficulties in assignment are expected for the remaining six signals which have β + γ isotope effects, especially as they group into signals with similar chemical shifts _i.e._ C-1 and C-6 between 62.7 and 63.2 ppm and C-3, C-5, C-2', C-4' between 70.2 and 72.6 ppm. Based on expected chemical shifts and magnitudes of isotope effects C-4' (_ca_ 70 ppm; 104, 29 ppb) can be differen-tiated from C-3 (_ca_ 70 ppm; 95 ppb and very small) whereas differences are not so clear-cut for C-2' (_ca_ 72 ppm; 84, 20 ppb) and C-5 (_ca_ 72 ppm; 95, 19 ppb) and for C-1 (_ca_ 63 ppm; 114, 27 ppb) and C-6 (_ca_ 63 ppm; 114, 24 ppb). It can be seen that analysis of the SIMPLE ^{13}C NMR spectrum of maltitol (Table 1) enables 8 out of 12 signals to be assigned unequivocally by inspection and the remaining 4 signals are differentiated into 2 pairs of signals with similar characteristics _i.e_ C-2'/C-5 and C-1/C-6. Chemical shifts of maltitol are summarised in Table 1 and magnitudes of β and γ isotope effects are summarised in Table 2.

The ^{13}C NMR assignment of maltitol in aqueous solution can be determined by the DIS NMR method using the magnitudes of isotope effects determined by the SIMPLE NMR

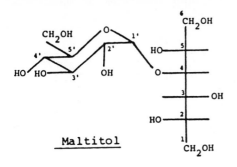

Isotope effects

Nucleus	Expected
C-1	$\beta+\gamma$
C-2	$\beta+2\gamma$
C-3	$\beta+\gamma$
C-4	2γ
C-5	$\beta+\gamma$
C-6	$\beta+\gamma$
C-1'	γ
C-2'	$\beta+\gamma$
C-3'	$\beta+2\gamma$
C-4'	$\beta+\gamma$
C-5'	2γ
C-6'	β

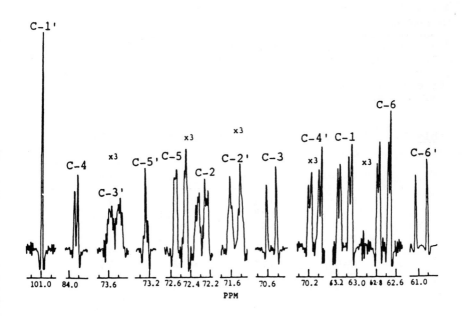

Figure 4.   100MHz ^{13}C SIMPLE NMR spectrum of maltitol in $(CD_3)_2SO$ solution 310K.   (Adapted from ref. 15 and reproduced by kind permission of Magn. Reson. in Chem.)

Table 1. Assignment of the ^{13}C-NMR spectrum of Maltitol.[a]

Nucleus	H$_2$O	D$_2$O	DIS	(CD$_3$)$_2$SO[b]
C-1	65.780	65.638	0.142	63.178[d]
C-2	74.475	74.357	0.118	72.345
C-3	73.340	73.249	0.091	70.626
C-4	84.604	84.604	[c]	83.943
C-5	75.603	75.493	0.110	72.559[d]
C-6	65.280	65.142	0.138	62.780[d]
C-1'	103.264	103.264	[c]	100.998
C-2'	74.519	74.413	0.106	71.626[d]
C-3'	75.847	75.690	0.157	73.610
C-4'	72.344	72.235	0.109	70.213
C-5'	75.307	75.307	[c]	73.294
C-6'	63.394	63.274	0.120	61.021

[a] 100MHz spectra were obtained at 310K. Chemical shifts
in ppm are quoted relative to dioxan = 69.371 for aqueous
solutions and to (CD$_3$)$_2$SO = 39.552 for (CD$_3$)$_2$SO solutions.
[b]The relative order of closely spaced peaks (C-5/C-5' and
C-2/C-2') is different between (CD$_3$)$_2$SO and aqueous
solutions and has been checked by a mixed solvent study.
[c]Effect too small to observe. [d]Not assigned by
SIMPLE NMR.

Table 2. ^{13}C SIMPLE NMR Isotope effects of Maltitol [a]

Nucleus	β	γ	(β+γ)
C-1	114	27	141
C-2	88	17,17	122
C-3	95	[b]	95[c]
C-4	–	[b],35	35[c]
C-5	95	19	114
C-6	114	24	138
C-1'	–	[b]	[b]
C-2'	84	20	104
C-3'	90	44, 22	156
C-4'	104	29	133
C-5'	–	22, 22	44
C-6'	116	–	116

[a] All isotope effects are negative and magnitudes are
quoted in ppb (i.e. 10^{-3} ppm), [b] Effect too small to
observe. [c] Minimum magnitude as one γ effect is very small.

method. The 50 MHz ^{13}C DIS NMR spectrum of maltitol
(Figure 5) displays two signals for each carbon atom, one
for the sample in D_2O and one for the sample in H_2O
solution. Three resonances (later identified as C-1', C-4
and C-5') appear as singlets because the magnitudes of the
isotope effects are small (ca 10-40 ppb) and signals from
two resonances overlap (later identified as C-2 and C-2')
though from the normal ^{13}C NMR spectrum it is found that
their chemical shifts are separated by 0.05 ppm. Most of
the maltitol signals can be assigned in D_2O and H_2O
solutions by comparison of the size of the DIS effect with
the sum of the separate isotope effects in the equivalent
SIMPLE NMR multiplet (Tables 1 and 2), after correcting
for the bulk magnetisation effect between the samples in
the two tubes.[14] Of the two sets of closely-spaced
signals one set (C-3', C-5' and C-5) can be readily
identified from the magnitudes of the DIS effects (157,
very small and 100 ppb, respectively) whereas
differentiation of the other set (C-2 and C-2') is less
clear cut because the magnitudes of the DIS effects are
similar (118 v 106 ppb); it is expected that the larger
isotope effect is given by the C-2 resonance and the
smaller effect given by C-2', an assignment which was
confirmed by the 2D-INADEQUATE spectrum.[15] Although the
DIS NMR spectrum was able to resolve the ambiguity in
assignment of C-2' and C-5 (not possible by SIMPLE NMR),
it still needed the 2D-INADEQUATE NMR spectrum to resolve
the ambiguity in assignment of the C-1 and C-6 signals.[15]

It should be noted that there are relatively large
changes in chemical shifts of maltitol in DMSO and aqueous
solutions (ca 2-3 ppm) except for C-4(ca 0.7 ppm), the
atom to which the glycosidic bond is attached. With such
large changes in chemical shifts it is difficult to assign
signals in different solvents by analogy, as shown in the
present work on maltitol where there is cross-over of two
sets of signals (C-2/C-2' and C-5/C-5') between DMSO and
D_2O solutions (Table 1); the cross-over of chemical
shifts was confirmed by a mixed solvent study.

5. APPLICATION OF SIMPLE ^{1}H NMR TO STUDY HYDROGEN-
    BONDING

Secondary isotope effects have been observed by ^{1}H NMR
spectroscopy in such cases as cyclodextrin[5], sucrose[5,6],
glucose[22] and maltose[22] where intramolecular hydrogen
bonding occurs between groups that both contain
exchangeable atoms,
       i.e. X-H..... Y-H   v.   X-D .... Y-H

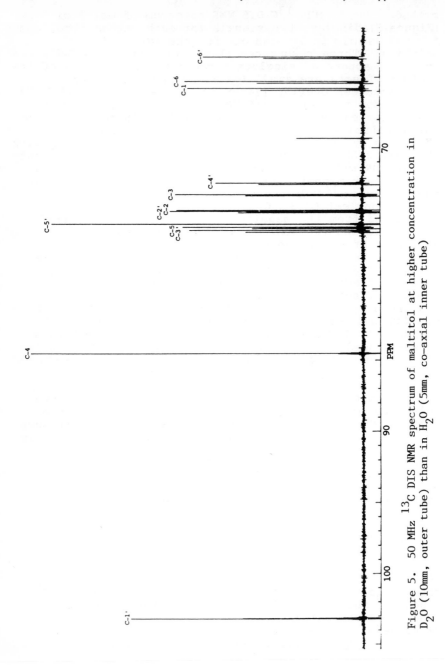

Figure 5.   50 MHz ^{13}C DIS NMR spectrum of maltitol at higher concentration in D_2O (10mm, outer tube) than in H_2O (5mm, co-axial inner tube)

in which group Y-H is observed; for carbohydrates
X=Y=oxygen atom. Analysis of these isotope effects
provides information on the presence of hydrogen bonding,
gives an indication of the strength of the hydrogen bond
and the predominant donor and acceptor groups, enables
hydrogen-bond networks to be observed, and shows that such
hydrogen bonding is cooperative *i.e.* as more hydroxy
groups take part in the hydrogen-bond network the hydrogen
bonds in the network become stronger.[16] SIMPLE ^{1}H NMR
measurements of sucrose have also shown that there are two
inter-residue hydrogen bonds in competition for one
acceptor group and the competitive equilibrium favours one
hydrogen-bonded conformational form.[5] All these aspects of
hydrogen bonding will be exemplified in the present work
using the spectrum of a sucrose derivative.

Origins of Isotope effects on Hydroxyl Proton Resonances

One isotope effect When two hydroxyl groups are
involved in a hydrogen-bond interaction (Scheme 1, figure
6), signals for 1-OH correspond to 2-OH isotopomers (i) and
(ii) (labelled H and D, respectively) whereas 2-OH signals
correspond to 1-OH isotopomers (i) and (iii). The relative
intensities (I) of isotopomer hydroxy proton signals are
directly proportional to the OH:OD ratio. Negative isotope
effects correspond to signals shifted to lower frequencies
on substitution with the heavier isotope, whereas positive
isotope effects correspond to signals shifted to higher
frequencies. From measurements on cyclodextrin[5] and
sucrose[5,6], it was found that the donor hydroxyl group
exhibits a normal negative isotope effect whereas the
acceptor hydroxyl group exhibits the unusual positive
isotope effect, though our recent measurements indicate
that this is not a necessary condition.[23]

Two isotope effects. The case where the hydroxy group
being considered (2-OH) is hydrogen-bonded to two different
hydroxy groups (1- and 3-OH) is shown diagrammatically in
Scheme 2 together with the four possible isotopomers (HH,
HD, DH, and DD, reflecting the state of deuteriation of 1-
and 3-OH). The relative intensities (I) of the isotope-
shifted resonances of the 2-OH signal corresponding to each
species depends on the OH:OD deuteriation ratio as shown in
Scheme 2 (Figure 6) in terms of the relative population (p)
of the protio and deuterio forms.

Ignoring the effect of proton spin-coupling constants,
which effectively doubles the pattern of resonances, the
appearance of the 2-OH resonances for different signs and
magnitudes of the two isotope effects is shown in Figure 6

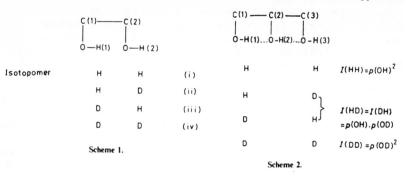

Scheme 1.

Scheme 2.

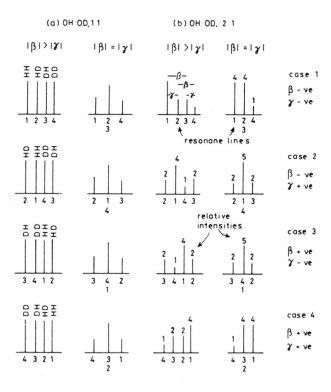

Figure 6. Dependence of the number of lines and their relative intensities on the signs and magnitudes of two isotope effects (labelled β and γ) at OH:OD ratios of (a) 1:1 and (b) 2:1. (Adapted from ref. 22 by kind permission of J. Chem. Soc.)

at two deuteration ratios. If the magnitudes of the two
isotope effects are different, four lines are expected
corresponding to the different isotopomers, whereas three
lines are observed if the magnitudes of the two isotope
effects are equal. Designation of lines in each multiplet
to the four isotopomers depends on the relative signs of
the isotope effects. Resonance signals with similar
distributions of intensities are expected for isotope
effects with either the same signs (Cases 1 and 4) or
different signs (Cases 2 and 3).

Hydrogen bonding in 6,6'-dichloro-6,6'-dideoxysucrose.

Part of the 500 MHz SIMPLE ^{1}H NMR spectrum of 6,6'-
dichloro-6,6'-dideoxysucrose (Figure 7) shows that isotope
effects are observed on most hydroxy signals in addition to
the vicinal proton coupling to the corresponding methine or
methylene protons. Analysis of the spectrum at different
OH:OD ratios enables the signs and magnitudes of isotope
effects to be determined and shows that no isotope effects
are observed for OH4', one isotope effect is observed for
OH1' (-4.6 ppb) and OH4 (small effect observed as line
broadening), two isotope effects of opposite signs are
observed for OH3' (±1.8 ppb) whereas three isotope effects
with different magnitudes are observed for OH2 (+7.9,+2.7
and |0.8| ppb) and with similar magnitudes for OH3 (+16,
+16, -16 ppb). The magnitudes and signs of isotope effects
are summarised in Table 3 together with those for sucrose
and for 1'- and 3'- substituted sucrose derivatives needed
to explain them in terms of hydrogen bonding. In SIMPLE [1]H
NMR spectroscopy each separate isotope effect on a hydroxyl
group corresponds to a different hydrogen bond to that
hydroxyl group.

Inter-residue intramolecular hydrogen bonding. The
large positive isotope effect on OH2 (+7.9 ppb, acceptor)
and the negative isotope effect on OH1'(-4.6 ppb, donor)
correspond to the presence of the OH1'....O2 inter-residue
hydrogen bond previously found in [^{2}H$_6$]Me$_2$SO solutions of
sucrose[5,6] and 3'-sucrose derivatives (3,3',4',6'-tetra-O-
acetylsucrose and 3',6'-di-O-benzoylsucrose).[24] The
medium-sized isotope effect on OH2(+2.7 ppb, acceptor) and
the isotope effect on OH3'(-1.8 ppb) correspond to the
presence of the novel inter-residue OH3'....O2 hydrogen
bond recently observed for 1'-sucrose derivatives.[16]
Observation of these isotope effects is consistent with
either a bifurcated hydrogen bond in which OH2
simultaneously acts as acceptor for both the OH1' and OH3'
donors (excluded because of unfavourable steric
interactions as shown by molecular models) or, more

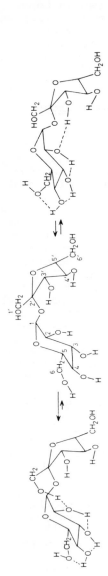

Figure 8. Conformational equilibrium of sucrose showing two inter-residue hydrogen-bonded forms in solution and the hydrogen-bond network in the glucose residue stabilised by the inter-residue hydrogen bonds. Hydrogen bonds in the fructose residue have not been characterised completely (Reproduced from ref. 5 by permission of J. Chem. Soc.)

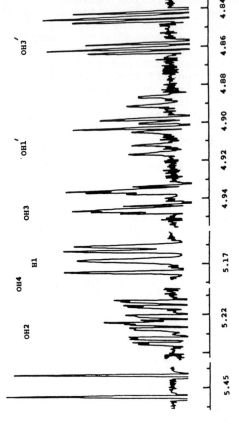

Figure 7. 500 MHz NMR spectrum of the hydroxyl proton resonances of 6,6'-dichloro-6,6'-dideoxy-sucrose in $DMSO\text{-}d_6$ solution at OH:OD ca 1:1

likely, the two inter-residue hydrogen bonds exist in competitive equilibrium, as shown in Figure 8. Assuming that the magnitudes of isotope shifts reflect the relative 'strengths' of hydrogen bonds, it is found that the equilibrium favours the OH1'.... O2 compared to the OH3'...O2 hydrogen bond in 6,6'-dichloro-6,6'-dideoxy-sucrose as well as in sucrose (Table 3).

Intramolecular hydrogen-bond networks. The smaller isotope effects observed for most hydroxy signals correspond to very weak hydrogen bonds between neighbouring hydroxy groups which are stabilised by the presence of a relatively strong and predominantly unidirectional inter-residue hydrogen bond; such isotope effects are not observed for the corresponding hydroxy resonances of the monomer units of sucrose, methyl-α-$\underline{D}$-glucopyranoside and methyl-β-$\underline{D}$-fructofuranoside.

The hydrogen-bond network in the glucose residue of 6,6'-dichloro-6,6'-dideoxysucrose is smaller than that observed previously for 1'- and for 3'-sucrose derivatives,[16,24] as shown in the separate hydrogen-bond schemes for sucrose in Figure 8. The magnitudes of the isotope effects in the glucose residue generally decrease in the series OH2>OH3>OH4>OH6 corresponding to progressively weaker hydrogen bonding at distances further from the intramolecular hydrogen-bond. The effect can be seen for both the 1'- and 3'-substituted sucrose derivatives as well as in sucrose and 6,6'-dichloro-6,6'-dideoxysucrose (Table 3). For example, in 3',6'-di-$\underline{O}$-benzoylsucrose the OH2 ....OH1' intramolecular hydrogen bond stabilises a hydrogen-bond network in which OH3 acts as both donor and acceptor hydroxyl group ($\pm$2.2 ppb) to OH2 (line broadening) and OH4 (line broadening) with even weaker hydrogen bonding from OH4 to OH6 only observed as line-broadening. Due to substitution of OH6 in 6,6'-dichloro-6',6'-dideoxy-sucrose the hydrogen-bond network stops at OH4 with isotope effects being observed for OH2 ($|0.8|$ ppb) and OH3 ($\pm$1.6 ppb) and a small effect on OH4 as a result of line-broadening.

i.e. ---OH$_2$ .... OH3 ....OH4

Only two isotope effects of opposite sign are expected for OH3 acting both as donor and acceptor in a hydrogen-bond network whereas three are observed for both sucrose[5] and 6,6'-dichloro-6,6'-dideoxysucrose (Figure 7); one possible explanation is that the separate inter-residue intra-molecular hydrogen bonds stabilise separate hydrogen-bond networks (as in Figure 8) but that separate isotope

Table 3. Magnitudes and signs of isotope effects ($\times 10^{-4}$ ppm)[a]

	Fructose residue				Glucose Residue		
	OH4'	OH3'	OH1'	OH2	OH3	OH4	OH6
1'-substituted sucrose derivatives[b]							
I		-13	-	+30			
II		-14	-	+31			
	br	+14		br	br		
III		-21	-	+40	-13	br	br
	-15	+21		br	+13		
3'-substituted sucrose derivatives[c]							
3,3',4',6'-tetra-O-acetyl sucrose		-30		+117			
3',6'-di-O-benzoyl sucrose		-63		+125	-22	br	(br)
				(br)	+22		
Sucrose [d]							
	-12	-40		+70	+15	br	br
	+12			+32	+15		
	(+br)			(br)	-15		
6,6'-dichloro-	+18	-46		+79	+16	br	-
6,6'-dideoxy-	-18			+27	+16		
sucrose				|8|	-16		

a 500MHz ^{1}H NMR measurements in [[2]H$_6$]Me$_2$SO solution at a data resolution of <0.1 Hz per point unless otherwise stated. Error limits $\pm$ 2x10^{-4}ppm. Br= broad signal signifying isotope effect ($\leqslant$ 10 x 10^{-4} ppm) too small to be measured. b Ref.16. I, 3-O-methyl-1,6-dichloro-1,6-dideoxy-β-D-fructofuranosyl 4-chloro-4-deoxy-α-D-galacto pyranoside; II, 1,6-dichloro-1,6-dideoxy-β-D-fructo-furanosyl 4-chloro-4-deoxy-α-D-galactopyranoside; III, 1'6'-dichloro-1',6'-dideoxysucrose. c. Ref.24, d Ref.5

effects are only observed for OH3 which normally exhibits the largest isotope effects in the network.

Hydrogen bonding in the fructofuranoside residue is indicated by the two isotope effects on OH3' of similar magnitudes and of opposite signs ($\pm$1.8 ppb) suggesting that OH3' acts as both donor and acceptor of hydrogen bonds; one isotope effect corresponds to OH3'....OH2 inter-residue hydrogen bonding and the other to OH4'....OH3' intra-residue hydrogen bonding as found previously for 1'-substituted sucrose derivatives.[16]. The presence of this hydrogen bond in the fructofuranose residue is consistent with the $^3T_4$ conformation of the fructofuranose ring (given by large magnitudes of J_{34} and J_{45} $\underline{ca}$ 8-9Hz) and the predominant H3'-C3'-O-H $\underline{anti}$ conformation for the OH3' group (given by the large magnitude of $J_{HCOH3'}$ $\underline{ca}$ 8Hz).[16]

<u>Cooperative hydrogen bonds</u> The magnitudes of isotope effects observed for hydroxyl groups depend not only on the hydrogen-bond type and its intrinsic strength but also on the relative populations of hydroxyl group conformations suitable for hydrogen bonding to occur. The weak intra-residue hydrogen bonding between neighbouring hydroxyl groups in the glucose and fructose residues of sucrose and its derivatives is consistent with the fact that there is considerable flexibility of the hydroxyl groups as shown by comparison of their vicinal coupling constants (range 4.6-6.0 Hz) that are similar to those observed for the constituent monomers and to those observed for "free" rotation in alcohols (5.3 Hz)[25]; a notable exception is OH3' (7.6-9.3 Hz)[5,16] which corresponds to approximately 50-80% of the <u>anti</u> conformer that is necessary to form the OH3'...O2 intramolecular hydrogen bond.

Despite the limitations on quantitation by the conformational flexibility of the molecule, it is expected that magnitudes of isotope effects reflect the relative "strengths" of hydrogen bonds. Analysis of the results in Table 3 shows that, as more hydroxyl groups take part in the hydrogen-bond network, the relative "strengths" of the inter-residue hydrogen bond increases and, concomitantly, the relative "strength" of the hydrogen bonds in the network increases <u>i.e.</u> the hydrogen bonding is a cooperative process. For example, in the 1'-substituted sucrose derivatives as more hydroxyl groups are involved in the network (compounds I-III, Table 3) the strength of the OH3'...OH2 hydrogen bond increases as shown by the increase in the isotope effects on both OH3' and OH2 and hydrogen bonding in the hydrogen-bond network becomes stronger in both the glucose and fructose residues.[16]

6. SUMMARY

Applications of the use of H/D isotope effects on nuclear shielding shielding are summarised in Table 4 which includes examples from work already published and some potential applications.

Table 4. Applications of NMR Isotope effects

Application	Nucleus	Example	Ref.
1. Structural analysis			
(DIS)	^{13}C	Carbohydrates	8-10
(SIMPLE)		Carbohydrates	11-13
(SIMPLE + DIS)		Glucodisaccharides	14
(SIMPLE + DIS)		Maltitol	15, this work

2. Hydrogen bonding (SIMPLE NMR)

Carbohydrates	$^1H, ^{13}C$	C-O.... H-O- H	Sucrose	5,6
			α1->4 glycosides	22
			6,6'dichloro- 6,6'-dideoxysucrose	this work
Steroids			16-Epiestriol	23
Carbohydrates	^{13}C	C>O ...H-O- C	β1->4 glycosides	-
Steroids etc.	^{13}C	>C=O ... H-O-		27
Peptides etc.	^{13}C	>C=O ... H-N<	inter/intra	-
Nucleic acids	^{13}C	>C=O ... H-N-	(2xG-C,1xA-T)	26
	$^{13}C, ^{15}N$	C>N ... H-N< C	(1xG-C, 1xA-T)	

3. Interactions of water with molecules (slow exchange condition)

SIMPLE	X,Y	i.e.	X...H - O - H...Y	-
			X...H - O - D...Y	-
			X...D - O - H...Y	-
			X...D - O - D...Y	-

The use of SIMPLE ^{13}C NMR for structural analysis is well established[11-14] and, when used in conjunction with DIS NMR[14,15], becomes a very powerful method of assignment in aprotic solvents and in aqueous solution as shown in the present work for maltitol.

The use of SIMPLE ^1H NMR provides experimental evidence for a wide range of hydrogen bonds, from the relatively strong inter-residue hydrogen bond to the weaker intra-residue hydrogen bonds in the same molecule, as shown in the present work for the sucrose derivative 6,6'-dichloro-

6,6'-dideoxysucrose. An important feature of the SIMPLE NMR method of analysis is that measurements of very weak intramolecular hydrogen bonding of hydroxy groups can be made even though they may be participating in relatively strong intermolecular (e.g. solvent) hydrogen bonding at the same time. Examples of intramolecular hydrogen bonding have already been observed by the isotope shift method for carbohydrates[5,6,17,22] and steroids.[23] ^{13}C NMR spectroscopy has been used to study intramolecular hydrogen bonding in $\alpha 1 \rightarrow 4$ glucosides[11] and for intra- and inter-molecular hydrogen bonding for nucleic acids and other molecules.[26] Novel long-range ^{1}H and ^{13}C NMR isotope effects transmitted via hydrogen bonds in the macrolide antibiotic Bafilomycin A, provides evidence for a directed hydrogen-bond network involving both hydroxyl and carboxyl groups (i.e. OH...OH...O=C)[27] and shows that the SIMPLE NMR method has the potential to study a range of hydrogen-bond types. The method also has the potential to investigate hydrogen-bond interactions of water with other molecules as long as the slow exchange condition is maintained; we await publication of the phenomenon with anticipation.

Acknowledgements We thank the S.E.R.C. for a post-doctoral research assistantship (to J.C.C.), the A.F.R.C. for a studentship (to R.E.H., in association with Tate and Lyle) and the M.R.C. for access to 500 MHz ^{1}H NMR facilities (N.I.M.R., London) and NMR computing facilities (Birkbeck College). We also thank the University of London for access to 400 MHz NMR facilities and Tate and Lyle for providing the sample of 6,6'-dichloro-6,6'-dideoxysucrose.

REFERENCES
1. R.A. Bernheim and H. Batiz-Hernandez, <u>Progress in Nucl. Magn. Reson. Spectroscopy</u>, 1967, <u>3</u>, 63.
2. P.E. Hansen, <u>Ann. Rep. in NMR Spectroscopy</u>, 1983, <u>15</u>, 106.
3. L.J. Altmann, D. Laungani, G. Gunnarsson, H. Wennerström, and S. Forsen, <u>J. Am. Chem. Soc.</u>, 1978, <u>100</u>, 8264.
4. P.E. Hansen, <u>Magn. Reson. in Chemistry</u>, 1987, <u>24</u>, 903
5. J.C. Christofides and D.B. Davies, <u>J. Chem. Soc., Chem. Commun.</u>, 1982, 560; <u>ibid</u>. 1985, 1533
6. R.U. Lemieux and K. Bock, <u>Jpn. J. Antibiot.</u> Supplement XXXII, 1979, S163; K. Bock and R.U. Lemieux, <u>Carbohydr. Res.</u>, 1982, <u>100</u>, 63.

7. C. Jameson, <u>Bull. Magn. Reson.</u>, 1981, <u>3</u>, 3; C. Jameson
    and H.J. Osten, <u>J. Chem. Phys.</u>, 1984, <u>81</u>, 4300.
8. P.E. Pfeffer, K.M. Valentine, and F.W. Parrish, <u>J. Am.</u>
    <u>Chem. Soc.</u>, 1979, <u>101</u>, 1265.
9. P.E. Pfeffer, F.W. Parrish and J. Unruh, <u>Carbohydr.</u>
    <u>Res.</u>, 1980, 84, 13.
10. P.E. Pfeffer and K.B. Hicks, <u>Carbohydr. Res.</u>, 1982, <u>102</u>,
    11.
11. J.C. Christofides and D.B. Davies, <u>J. Am. Chem. Soc.</u>,
    1983, <u>105</u>, 5099.
12. J. Reuben, <u>J. Am.Chem. Soc.</u>, 1983, <u>105</u>, 3711; <u>ibid</u>.
    1984, <u>106</u>, 2461 and 6180; <u>ibid</u>, 1985, <u>107</u>, 1433 and
    1747; <u>ibid</u>. 1986, <u>108</u>, 1735; <u>ibid</u>. 1987, <u>109</u>, 316.
13. J.C. Christofides and D.B. Davies, <u>J. Chem. Soc.</u>, <u>Perkin</u>
    <u>Trans II</u>, 1984, 481.
14. R.E. Hoffman, J.C. Christofides, and D.B.Davies,
    <u>Carbohydr. Res.</u>, 1986, <u>153</u>, 1.
15. R.E. Hoffman and D.B. Davies, <u>Magn. Reson. Chem.</u> 1988.
    (accepted for publication)
16. J.C. Christofides, D.B. Davies, J.A. Martin and E.B.
    Rathbone, <u>J. Am. Chem. Soc.</u>, 1986, <u>108</u>, 5738.
17. J. Reuben, <u>J. Am. Chem. Soc.</u>, 1985, <u>107</u>, 5867.
18. T.H. Mareci and R. Freeman, <u>J. Magn. Reson.</u>, <u>48</u>, 158
    (1982).
19. D. L. Turner, <u>J. Magn. Reson.</u>, <u>49</u>, 175 (1982).
20. P. Colson, K.N. Slessor, H.J. Jennings, and I.C.P.
    Smith, <u>Can. J. Chem.</u>, <u>53</u>, 1030 (1975).
21. S.T. Angyal and R. LeFur, <u>Carbohydr. Res.</u>, <u>84</u>, 201
    (1980).
22. J.C. Christofides and D.B. Davies, <u>J. Chem. Soc.</u>,
    <u>Perkin Trans. II</u>, 1987, 97.
23. D.B. Davies, M.A. Kelly and C.J. Stilwood, unpublished
    results.
24. J.C. Christofides and D.B. Davies, <u>Magn. Reson. Chem.</u>,
    1985, <u>23</u>, 582.
25. R.R. Fraser, M. Kaufman, P. Morand and G. Govil, <u>Canad</u>
    <u>J. Chem.</u>, 1969, <u>47</u>, 403.
26. J. Reuben, <u>J. Am. Chem. Soc.</u>, 1987, <u>109</u>, 316.
27. J. Everett, <u>J. Chem. Soc.,Chem. Commun.</u>, 1987, 1878.

7
The Use of Stable Isotopes in Medicinal Chemistry

D. Halliday and G.N. Thompson

NUTRITION RESEARCH GROUP, MRC CLINICAL RESEARCH CENTRE, HARROW, HA1 3UJ, UK

1 INTRODUCTION

Stable isotopes have been employed increasingly as tracers over the last decade both to provide the clinician with the opportunity to broaden, in a quantitative manner, discrete areas of diagnosis and research, and the clinical chemist with definitive methodology for specific analyte analysis[1,2]. These non-radioactive 'heavy' isotopes contain one or more extra neutrons in the nucleus compared with their more abundant 'lighter' analogues. This mass difference coupled with the small changes in chemical and physical properties imparted to a molecule by inclusion of a heavy isotope have been exploited in the separation and production of simple compounds containing highly enriched stable isotopes by cryogenic distillation, thermal diffusion or chemical exchange techniques. Impetus in the application of stable isotopes for *in vivo* studies has come from an increased awareness of the possible harmful effects in the use of radionuclides, and a realisation of several positive advantages conferred by the use of stable isotopes in their own right – certain elements of clinical importance (especially nitrogen) lack a useable radionuclide equivalent; use of a 'cocktail' of stable isotopes permits a range of studies to be performed in the same patient simultaneously and, within specific constraints, serial studies can be performed in the same patient.

More practical events that have influenced the useage of these metabolic probes include the increasing range of stable isotope labelled compounds that are currently available and, not least, recent significant developments in mass spectrometric and sample preparation techniques required for quantitative stable isotope analysis. This

chapter presents an overview of selected clinical applications of stable isotopes which is biased towards the more
recent literature and towards our personal interests
involving the use of isotope ratio and gas chromato-
graphy/mass spectrometry or both.

Dietary intake and toxicity

Stable isotopes of clinical interest occur naturally
at the following levels: ^{2}H, 0.015%; ^{13}C, 1.108%; ^{15}N,
0.366%; ^{17}O, 0.038%; ^{18}O, 0.204%; ^{34}S, 4.215%. Predict-
ably, any adverse biological effects attributable to
stable isotope administration above natural abundance will
be greater for deuterium in replacing hydrogen than for
isotopes of the higher elements where the mass differences
are much smaller and physical and chemical properties are
almost identical. This fact is dramatically demonstrated
by the exposure of tadpoles, selected aquarium fish, flat
worms or paramecia to both 90% and 30% deuterated water[3].
At the higher enrichment all species died within 3 hours,
but all survived at the 30% enrichment level. The first
mammalian study to document the effect of deuterium
indicated that administration of 0.66 g of 100% D_2O to an
adult mouse produced marked signs of intoxication,
although the animal survived[4]. The maximum amount of
deuterium likely to be administered as a tracer organic
molecule in a typical stable isotope turnover study is
about 7 mg, or about 0.00005% of the whole body hydrogen
pool, an insignificant addition to the natural ^{2}H abundance of 0.015% (based on a primed continuous infusion of
10μmol/kg/h of a ^{2}H$_6$ labelled compound over 4 hours in a
70 kg man). Comprehensive coverage of all aspects of the
biological effects of deuterium is available[5].

No adverse effects have been reported following mice
attaining 60% ^{13}C-labelling[6]. Further, three generations
of mice have been raised in an atmosphere and provided
with drinking water containing 90% ^{18}O$_2$ and H$_2$^{18}O respectively[7]. Throughout the study no adverse physiological or
biochemical effects were noted and reproduction proceeded
normally through each generation with no increased infant
mortality. Estimates of average total daily intakes of
commonly employed stable isotopes are presented in
Table 1.

TABLE 1: DAILY INTAKES OF MAJOR ELEMENTS AND 'HEAVY' ISOTOPES

Daily intake		Elemental composition(g)				Basis of calculation
Component	Weight-g	C	H	O	N	
Protein	100	50	7	23	16	C,50%;H,7%;O,23%;N,16%
Fat	120	92	14	14	–	$C_{18}H_{34}O_2$
Carbohydrate	320	128	21	171	–	$C_6H_{12}O_6$
Water	1500	–	167	1333	–	H_2O
Oxygen	535	–	–	535	–	O_2
Total element (g)		270	209	2076	16	

'Heavy' isotopes	^{13}C	^2H	^{17}O	^{18}O	^{15}N
– total (g)*	2.99	0.31	0.77	4.23	0.06
– mg/kg*	42.7	0.44	11.0	60.4	0.8

* calculated from natural abundances

2 INSTRUMENTATION

By far the most widely applied method for quantitation of stable isotope enrichment in biological samples is mass spectrometry. In common use are isotope ratio mass spectrometers (IRMS) and mass spectrometers coupled to a gas chromatograph (GCMS). Attributes of both systems are compared in Table 2. Both instruments work on the principle that heavier or more highly charged isotopes of a given compound will be influenced more by electromagnetic fields than their 'lighter' analogues (Figure 1). Compounds are in general presented to mass spectrometers as a gas, either as the gas chromatograph effluent (GCMS) or as a stable gas at room temperature (IRMS). The gas is then ionised by passing through an ionising source incorporating either a beam of electrons (electron impact ionisation, GCMS and IRMS) or a high pressure reagent gas, commonly methane or ammonia (chemical ionisation, GCMS only). The different types of ionisation can have markedly differing effects on the fragmentation pattern of a given molecule (Figure 2, page 183). The ions then pass a series of focussing lenses, after which they are subjected to an electric and/or magnetic field which deflects them according to their mass and charge. The ions then pass to a detector, where an electron current is generated, usually by the bombardment of the charged molecules against a

metal plate. Finally an amplification system multiplies
this voltage to a recordable signal. Because the deflection of a given ion depends on both its molecular weight
and charge, the detector location of a given current is
usually expressed in terms of molecular weight and charge,
the m/z ratio.

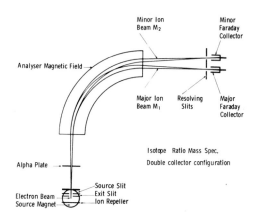

Figure 1: Typical configuration of an isotope ratio mass
spectrometer.

     Stable isotopes currently employed in clinical
studies occur naturally, and the measurement of interest
is the abundance of an isotope above the natural background. This is usually expressed as atoms percent excess
(APE) - the percentage of the compound of interest that is
isotopically labelled above natural background.

Table 2: FEATURES OF IRMS AND GCMS

	GCMS	IRMS
Sample introduction	Volatile derivative of complex molecule	$N_2, CO_2,$ H_2 or SO_2
Sample mass	>150	1-100
Sample size	Nanograms	Micrograms
APE range	>0.5%	0.0005-0.5%
Precision	+/-0.1%	+/-0.0002%

Isotope ratio mass spectrometry

IRMS is designed specifically for isotope ratio measurements of compounds in a gaseous form. Abundances of ^2H, ^{13}C, ^{15}N, ^{18}O and ^{34}S are most commonly measured as hydrogen, nitrogen, carbon dioxide and sulphur dioxide. Isotope ratio mass spectrometers possess low resolving power (100–200; 10% valley definition) and low mass range (m/z ratio; 2–100). Low resolution results from the width and position of beam resolving slits, from the low strength permanent or electro-magnet employed and from a short radius analyser (6–30cm) which restricts the ion beam path length. The instruments benefit from excellent precision (+/–0.0002%) and ability, therefore, to measure very small changes in the natural background. High sensitivity (0.2A ion current/mbar, about 0.0005 APE) results from 'z' focussing of the ion beams, the simultaneous collection of two or more isotope molecular ion beams ('light' and 'heavy') in separate buckets and the ability to rapidly alternate isotope ratio measurements of the unknown sample gas with a laboratory reference gas of known isotope composition.

Dual inlet systems permit the separate entry of reference and sample gases into the analyser. Each inlet system characteristically consists of small volume valves, variable volume bellows, metal capillaries and an automated change-over valve to facilitate the alternate entry of reference or sample gas into the analyser. The sample inlet can be connected to various on-line preparation systems or can itself consist of a fixed multiposition manifold, allowing for sequential automated sample admission. Automation is a major advantage, as manual IRMS is laborious, requiring about 1/2 h of technician time per sample.

Within the analyser the gas tight ion source operates in the electron impact mode at an electron energy of some 70eV and a trap current of 300–600μA. Molecular ions are extracted from the source under the influence of an accelerating potential (2–10kV) and the individual ion beams focussed into separate Faraday buckets. The radius of the path of the ion beam is proportional to the square root of the m/z ratio. Currents generated from the ion beams impinging on the collector plates are converted to a voltage, enhanced through a field effect transistor (FET) operational amplifier and then fed to a voltage frequency convertor. The output of this component is linearly proportional to the input voltage and the pulses originating from the 'heavy' and 'light' isotope molecular ion

beams are separately counted and the required ratio
obtained.

Gas chromatography/mass spectrometry

While GCMS was initially designed for identification
and analysis of complex organic molecules, software
advances have made possible the accurate determination of
the abundance of selected ions (selected ion monitoring or
SIM). The major disadvantage of GCMS is lack of sensitivity in measuring isotopic enrichment below about 0.1
APE. However, advantages in the sample size required
(nanogram range) and the ability to analyse complex
molecules make GCMS especially suited to the determination
of isotopic enrichments in plasma samples.

The technique of fast atom bombardment mass spectrometry (FAB-MS) allows the analysis of solid preparations
without prior derivatisation or gas chromatography, and
potentially provides the best method for the analysis of
non-volatile molecules[8,9]. The sample is usually placed
in a glycerol matrix, from which it is ionised directly.
Presently this method is limited by a high and sometimes
variable background which reduces its value for isotope
ratio measurements. Trace metal quantitation may also be
attempted using inductively-coupled-plasma mass spectrometry (ICP-MS) though application in terms of accuracy and
precision to biological matrices has yet to be thoroughly
tested[10].

Isotope ratio monitoring gas chromatography mass spectrometry (IRM-GCMS)

The reduced sample size required by modern isotope
ratio mass spectrometers has led to the ability to couple
these instruments with a gas chromatograph. The effluent
of the gas chromatograph is catalytically combusted to CO_2
or N_2 and the gas then admitted to the IRMS. This technique retains the gas chromatography selectivity while incorporating the IRMS sensitivity to isotope enrichment[11].

Thermal ionisation mass spectrometry

Analysis of metal ions with precision approaching
that of IRMS is possible by thermal ionisation mass
spectrometry (TIMS)[12]. Ions are generated in this method
by the application of the sample to a metal strip, to
which thermal energy is then applied. The high cost
(around 250,000 pounds) and intensive labour requirements
have restricted the application of this precise technique

in clinical studies. Nonetheless, several recent publications have indicated the use of TIMS to study kinetic aspects of calcium, iron, zinc and lead metabolism in human subjects[13-16].

3 SAMPLE PREPARATION

Isotope ratio mass spectrometry

Very precise measurement of the stable isotope content of a pure gas afforded by classical IRMS can only be as accurate and reproducible as the sample offered for analysis. Common to all sample preparation procedures for IRMS analysis is the need to maintain isotope continuity throughout the range of chosen reactions that span the intermediate steps occurring in the conversion of initial sample material to final gaseous product. Generally, lighter isotopes of an atom have a greater reactivity than their heavier equivalent, so that the product or products formed initially in a reaction (enzymatic or chemical) will be enriched with respect to the 'lighter' isotope. Precursor-product isotope integrity can only be achieved therefore if each step within a complete gas sample preparation can proceed or be driven to completion. Gaseous hydrogen required for deuterium assay by IRMS can be obtained by reduction of the aqueous phase derived from body fluids (urine, saliva, serum) following vacuum distillation. Water reduction may be performed directly on-line to the mass spectrometer, though it is normally more convenient to operate a stand-alone dedicated preparation line employing zinc (400°C)[17,18] which reduces water:

$$H_2O + Zn \longrightarrow ZnO + H_2$$

as does magnesium (5-600°C)[19]:

$$H_2O + Mg \longrightarrow MgO + H_2$$

or uranium (6-800°C)[20,21]:

$$2H_2O + U \longrightarrow UO_2 + 2H_2$$

Hydrogen in organic molecules of interest derived from chemical studies where deuterium label has been employed can be converted to hydrogen gas either directly by combustion over carbon, nickel or platinum[22,23] or by combustion to water as an initial step, over cupric oxide (static[24] or dynamic systems[25]) and thence effecting the

separate reduction of water to gaseous hydrogen as outlined above. Hydrogen preparation for IRMS analysis has
recently been reviewed in detail[26].

[13]C-labelled compounds have become increasingly
available in the last 5 years. Measurements of clinical
interest relating to IRMS centre around the incorporation
of ^{13}C label into specific compounds or the rate of oxidation of administered substrates. In the former case,
provided exact knowledge of the site of the label is not
required, and that the label will not be excessively
diluted by unlabelled carbon from the molecule skeleton,
the purified compound can be combusted in oxygen and the
resulting CO_2 cryogenically recovered. Recently elemental
analysers (combustion systems) have been coupled on-line
to IRMS to provide both quantitative elemental and isotope
enrichment data. A further extension of this system is to
utilise the separatory power of gas chromatography to
achieve initial separation of the compound of interest
which can then be passed directly to the combustion
chamber as referred to above. Several systems to tackle
these analytical approaches have been described[12,27-29].

Methods for the collection of respiratory [13]CO_2 to
monitor oxidation rates include the direct transfer of
breath into evacuated septum-capped tubes or greaseless
tapped bottles to avail CO_2 cryogenic separation[30] or the
precipitation of CO_2 as a carbonate and subsequent release
by action of 85-100% phosphoric acid under vacuum[31] prior
to isotope analysis. Both these approaches have been subjected to almost complete automation such that 3-4
complete analyses can be conducted per hour.

Isotope fractionation: Photosynthetic carbon
assimilation by green plants provides a mechanism for the
isotope fractionation of carbon such that both terrestrial
and marine plants have a lower ^{13}C/^{12}C ratio than their
respective carbon sources (atmospheric CO_2 and oceanic
carbonates). Positive ^{12}C discrimination occurs at 2
points within the overall photosynthetic process: first, a
kinetic effect occurring at the point of uptake of
atmospheric CO_2 into the plant leaf cytoplasm and secondly
during the actual fixation of the dissolved CO_2 giving
rise to an initial product. This initial stable organic
compound may be a 3-carbon acid, 3-phosphoglycerate
(formed from ribulose diphosphate) if photosynthetic CO_2
fixation is via the Calvin-Benson (C_3) cycle or the 4-
carbon acids oxaloacetate, malate and aspartate if the
Hatch-Slack (C_4) cycle operates. Fixation of atmospheric
CO_2 to crassalacean acid represents a third pathway found

almost exclusively in succulent plants of arid regions. Examples of C_3 plants are sugar beet, rice and potato while sugar cane, corn and sorghum are C_4 plant types.

Further metabolic processes within the plant introduce additional carbon-isotope fractionation whereby in general lipid is 'lighter' than protein which in turn is 'lighter' than carbohydrate within a given group. The relevance of these observations to the present discussion relates to the fact that as man is at the end of a lengthy food chain, having its origins in photosynthetic plants, observed variations in the carbon isotope composition of the whole body, individual tissues and end products may not surprisingly be attributed to dietary intake. This concept assumes importance when quantitating the rate of oxidation of an administered ^{13}C-labelled compound whose ^{13}C content will undergo vast dilution before expulsion at the lung surface as metabolic CO_2 and in fact may produce a signal scarcely detectable above the inherent variations outlined above.

^{15}N preparation: Preparation of nitrogen gas for ^{15}N isotope analysis generally employs a Kjeldahl digestion stage performed on the separated and purified nitrogenous compound of interest. The ammonium salt produced is then oxidised by the hypobromite reaction to generate molecular nitrogen. Final oxidation can be adapted to allow batch handling or some automation of the reaction[32-34]. An alternative approach is to Dumas combust the nitrogenous compound in an elemental analyser, again coupled directly to the IRMS. Typical current analysis by this mode provides a precision in the range of 0.0005-0.001 atom % ^{15}N at natural abundance levels on 5 μg N within 5 minutes[35-37].

^{18}O and $^{2}H_2{}^{18}O$ preparation: Clinical studies involving the use of ^{18}O-label and assay by IRMS almost exclusively involve the use of $H_2{}^{18}O$ and $^{2}H_2{}^{18}O$ for total body water and energy expenditure estimates respectively (see below). In the former case the oxygen can be sampled as $C^{16}O^{18}O$ in expired air and measured as the m/z 46/44 ratio as indicated previously. In the latter case a commercial IRMS is now available to simultaneously measure ^{2}H and ^{18}O (as water vapour)[38,39]. Other possibilities exist including an equilibration technique[40] and the use of guanidine hydrochloride[41] to quantitate ^{18}O in samples generated from clinical studies.

^{34}S preparation: ^{34}S has as yet found little use in clinical or biological studies but its increasing avail-

ability suggests that this situation may change in the
near future. Analysis of [34]S by IRMS, in the context of
geochemical studies is mostly as sulphur dioxide though
the hexaflouride has also been used[42]. Sulphides can be
oxidised directly to SO_2 for isotope analysis and
sulphates can be thermally decomposed to yield the
dioxide. Alternatively both sulphide and sulphate can be
converted to hydrogen sulphide, the sulphur precipitated
as silver sulphide and this then reproducibly oxidised to
produce SO_2 for isotope analysis.

Gas chromatography/mass spectrometry

The requirements for GCMS sample admission are
similar to those for gas chromatography alone. The sample
usually undergoes one or more purification steps followed
by derivatisation of the compound of interest to form an
intermediate that is volatile and stable in the
temperature range 120–130°C. In addition, however, the
derivative must provide abundant ions of sufficiently high
m/z ratio under either chemical or electron impact ionisation to allow analysis of those ions without interference.
This generally requires a major ion fragment of m/z ratio
>150. In addition the derivatisation procedure must
preserve the labelled atom(s) within the major fragment to
be monitored. A derivatisation procedure which appears to
be convenient for the majority of molecules used in tracer
studies is the formation of the tertiary–butyl–dimethyl–
silyl (TBDMS) derivative described by Schwenk et al[43]. The
structure of this derivative, and its differing fragmentation patterns under chemical and electron impact ionisation is shown in Figure 2.

4 CLINICAL STUDY MODELS

The mathematical principles of clinical stable
isotope tracer studies are based on those for radioactive
tracers. Stable isotope tracers may be administered by
bolus dose (non–steady state model) or by constant infusion (steady state model). By complex mathematical analysis
of the isotopic decay of a labelled compound in plasma,
the former model may allow calculation of the pool
size(s), volume of distribution and turnover rate of the
compound[44]. This model is limited by the assumption that
the body pool of the tracee is homogenous and thoroughly
mixed, by the need for numerous accurately timed and
precisely analysed blood samples, and by the inability to
quantitate the oxidation of substrates to breath CO_2.

These limitations have led to the more widespread use of steady state models.

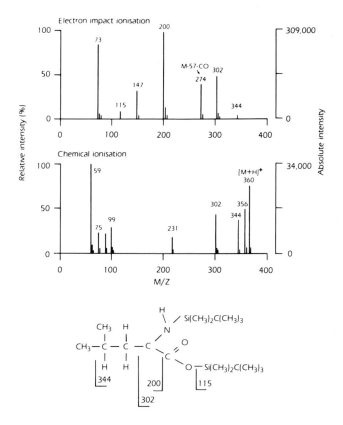

Figure 2: Fragmentation of TBDMS-leucine under chemical (CI) and electron impact (EI) ionisation. The peak heights represent the relative abundances of each m/z ratio under the different types of ionisation. The peak at m/z ratio 360 under CI represents the intact leucine derivative (molecular ion). No molecular ion is seen under EI. The fragmentation patterns of a given molecule are generally similar with different mass spectrometers, but the exact abundance of each ion can vary significantly.

While the steady state model still assumes a homogenous and thoroughly mixed body tracee pool, the
potential errors arising from physiological departure from
this ideal are relatively smaller than for the non-steady
state model. The steady state model does not allow calculation of tracee pool size or volume of distribution, but
in general can be expected to give a more precise measure
of turnover rate. Furthermore the oxidation of substrate
to breath CO_2 can be quantitated. The model (Figure 3) is
based on the concept that in the steady state the rate of
entry of tracee to the body pool must equal its rate of
departure from that pool. If a tracer of known isotopic
purity (APEi) is infused into the pool at a known,
constant rate (f μmol/min), then the dilution of the
tracer in the sampling pool (or the abundance of label in
that pool, APEp) will be in proportion to the rate of
appearance of unlabelled tracee (Q μmol/min). Thus:

$$Q = f . \left[\frac{APEi}{APEp} - 1 \right]$$

The component (-1) corrects for the contribution of
the rate of infusion of the tracer to the total turnover
of tracee. For most substrates the APEp is measured by gas
chromatography/mass spectrometry (GCMS).

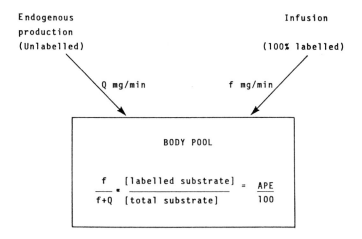

Figure 3:   Steady state model for the calculation of
substrate production rate using a stable isotope tracer
(see text).

This model has been successfully employed in the measurement of whole body protein turnover (Figure 4). The most commonly employed tracer for measuring protein turnover has been (^{13}C)leucine. In the fasting state, the only entry of leucine to the plasma pool arises from protein catabolism (C) while the only exits from the pool are to protein synthesis (S) and leucine oxidation (O). In the steady state, the total turnover, or flux, of the pool (Q) is equal to the entry and exit rates from the pool. Thus:

$$Q = C = S + O$$

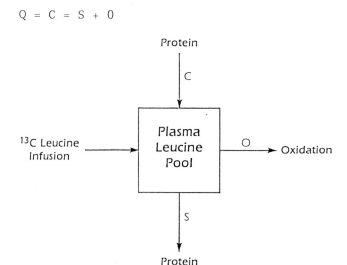

Figure 4: Model for the measurement of whole body protein turnover using (^{13}C)leucine (see text).

The rate of leucine oxidation can be calculated by measurement of the $^{13}CO_2$ enrichment of expired air by IRMS and the rate of total breath CO_2 production. Use of this type of model has permitted studies of amino acid metabolism as affected by varied dietary protein and energy intakes in normal and pathological states. Extensive literature is available regarding general amino acid metabolism studies[45,46] and studies relating to specific amino acids; lysine[47,48], glycine[49,50], valine[51,52], threonine[53], leucine[54-56] and alanine[57]. More complex analysis of the steady state model can allow calculation of the rate(s) of conversion between substrates and rates of tissue protein synthesis[47]. Fractional

protein synthesis rate calculations demand some approximation of the precursor pool enrichment in the absence of
the as yet practically unmeasurable appropriate amino acid
t-RNA enrichment. Various attempts to circumvent this
particular problem and measure individual protein and
forearm protein synthesis rates have been published[58-62].

The numerous adaptions of the non-steady state model
include the measurement of protein turnover by oral bolus
dose techniques[63], the measurement of total body water and
energy expenditure[64,65] (see below), the qualitative
analysis of enzyme defects (eg the detection of carrier
status for phenylketonuria[66]) and the various clinical
breath tests of gastrointestinal function (see below).

5 TECHNICAL DIFFICULTIES IN *IN VIVO* STABLE ISOTOPE STUDIES

Preparation of infusates

Stable isotope labelled compounds are becoming widely
available commercially, but in almost all cases the
preparations have not been cleared for use in clinical
studies. Official guidelines have not been established in
the UK and individual ethical committees generally assume
responsibility for approving the use of stable isotopes in
research studies. Fortunately, these compounds have been
in clinical use for more than 40 years, and, so long as
tracer doses are employed, no side effects attributable to
these labels have been reported. We consider the following precautions should be taken:

1. The unlabelled analogue of the tracer should be known
to be safe for administration and should preferably occur
naturally in humans.

2. The tracer should be obtained from a reputable source,
and should be stated by certificate to be of high chemical
purity. The chemical purity should be checked independ-
ently. Simple chromatography techniques are often
suitable, particularly for amino acids.

3. Solutions for intravenous infusion should be sterilised,
preferably by autoclaving. Unless the solution is to
be used immediately, it should be tested for sterility
by agar plate culture techniques. Solutions of compounds
which are not heat stable should be sterilised by passage

through a bacterial filter, preferably under laminar
flow conditions.

4. Solutions should be tested for pyrogens prior to use.
This should generally be undertaken when the solution has
been stored in its final form; freezing can release
pyrogens from solutions of palmitic acid complexed to
albumin[67].

5. Stability of the tracer in its final solution should
also be established. Assays should be performed for
exchange of label as well as for chemical degradation.

6. While the risk associated with the use of stable
isotopes appears to be very small indeed, this risk
should, as with any research study, be considered
appropriate in the context of the likely scientific value
of the study.

Label exchange

A stable isotope label may be lost at several stages:
during derivatisation of samples for mass spectrometric
analysis, during preparation of compounds for administration and *in vivo*. Fast atom bombardment (FAB) mass
spectrometry is a convenient method of measuring the
isotopic purity of a supplied compound without derivatisation, and can therefore be used to confirm loss of label
during derivatisation, storage or infusate preparation. *In
vivo* label exchange may be more difficult to detect, and
is likely to occur spontaneously with deuterium labels.
^{13}C and ^{15}N labels most commonly exchange only through
enzymic mechanisms. Stability of deuterium labels can be
supported through exposure to harsh conditions *in vitro*,
and, when multiply deuterated compounds are infused, by
monitoring multiple mass channels in plasma samples taken
in vivo.

Expired air CO_2 collection

Measurement of substrate oxidation relies on the accurate
estimation of the rate of expired air $^{13}CO_2$ production
arising from the labelled infusate. The limitations of
variability in baseline ^{13}C enrichment of expired air CO_2
and in the measurement of $^{13}CO_2$ enrichment by IRMS have
been discussed earlier. Two other considerations are
relevant: the rate of total expired air CO_2 production (V
CO_2), and the proportion of HCO_3^- in the body pool which
is actually excreted into expired air.

A number of techniques for measuring V CO_2 are available, including timed Douglas bag collections and ventilated hood systems[68]. No matter what technical accuracy these systems may have, however, they cannot account for physiological variations which can occur, for example, as a result of the subject hyperventilating or even talking during the collection. Thus, despite technical advances, the measurement of V CO_2 remains imprecise.

It is clear that not all CO_2 generated from a labelled substrate will be excreted in expired air. Exactly how much CO_2 is retained in the body HCO_3^- pool is, however, a matter of controversy, with estimates ranging from less than 10% up to 50%[72]. Furthermore, the percentage retention changes with physiological variations such as exercise[67] and fasting[69].

Sampling site and venous access

While most stable isotope studies are performed by infusing isotope into a superficial vein, the site from which blood samples are taken can influence results significantly[70]. During (^{13}C)leucine infusion, changes in (^{13}C)leucine enrichment occur between superficial venous (eg hand) and deep venous (eg cubital fossa) sites, and these changes differ from those which occur for (^{13}C)α-ketoisocaproic acid[71], with which leucine is in rapid equilibration intracellularly. Smaller changes also occur depending on the degree of 'arterialisation' of superficial venous samples. These changes result from the departure of the human plasma pool from the ideal of a thoroughly mixed pool into which isotope is infused and samples are taken. Difficulty in venous access can be particularly limiting to the performance of stable isotope studies in children. The inadvisability of varying the sampling site referred to above must be considered in the interpretation of children's studies.

6 APPLICATIONS

Stable isotope tracers can in general be applied in a similar manner to their radiolabelled analogues with the major advantage of ethical acceptability. There are, however, a number of distinct areas of research to which stable isotope labelled tracers alone can be applied. The applications summarised here demonstrate the specific advantages of stable isotope tracers, and present some of the recent advances that have come about through their use.

Substrate oxidation and breath tests

Fatty acid[72], amino acid[54] or carbohydrate[73,74] oxidation rates can be assessed semi-quantitatively by administration of the ^{13}C-labelled substrate of interest either as a bolus or constant infusion with subsequent monitoring of the $^{13}CO_2$ enrichment in expired air and the total CO_2 excretion rate. The product of enrichment and excretion rate will provide an indication of the rate of oxidation of substrate.

Non-invasive ^{13}C diagnostic breath tests may be regarded from a practical standpoint as a modification of standard ^{13}C oxidation studies. The theoretical basis of these tests (directed towards investigating gastrointestinal function) resides in the use of an appropriate ^{13}C-labelled substrate containing a 'target bond' which, following enzymatic cleavage, will release a small functional group destined to produce $^{13}CO_2$ as a metabolic end-product that will be excreted at the lung surface. Oral administration of the specific substrate dictates that the final appearance of $^{13}CO_2$ could reflect intestinal absorption, hepatic transformation, plasma distribution and/or end organ oxidation or any combination of these steps. For a given test, the highest diagnostic index will be obtained when the least number of rate-limiting steps are involved. Theoretical and practical considerations for the conduct of breath tests are well documented[75,76] (see above). Consideration has also been given to variations in ^{13}C intake (variations in natural abundance as discussed earlier) which would be superimposed on the breath test if performed in the fed state (which may be necessary in infant studies) and decrease the signal to noise ratio and thus the potential discriminatory power of the test[77,78]. ^{13}C breath tests provide a method for investigating ileal dysfunction and upper intestinal bacterial overgrowth using ^{13}C-glycholate[79], hepatic microsomal mixed oxidase activity (drug metabolism) in liver disease with ^{13}C-aminopyrine[80] and the severity of alcoholic cirrhosis using ^{13}C-galactose[81]. The use of ^{13}C-trioctanoin has provided positive discrimination of a variety of causes of fat malabsorption[82]. Recently a ^{13}C-urea breath test has been developed to investigate in more detail the known association between *Campylobacter pylori* and the presence of gastritis and of duodenal ulcer[83]. To date the prevalence, distribution and relapse after treatment (reappearance of *C.pylori*) have had to rely on the

invasive and relatively expensive endoscopic gastric
mucosal biopsy technique.

Multiple uses of labelled water (^2H_2O; $H_2{}^{18}O$ $^2H_2{}^{18}O$)

A 70 kg adult of 'normal' body composition contains
approximately 40 L of water to which additions in the form
of food or drink rapidly equilibrate. Ideally, isotopes of
hydrogen or oxygen should be used to monitor water movements within the body, to estimate total body water (TBW)
or to define water turnover rates. Deuterium label (^2H)
has been employed to monitor water movements within the
body relating to the formation of cerebrospinal fluid[84],
water uptake by the stomach[85], movement of water across
the skin[86] and to determine the rate of water equilibrium
between the body and dialysis fluid in an attempt to
improve the efficiency of peritoneal dialysis[87].

Accurate measurement of TBW assumes importance in
determining body composition and nutritional status in the
absence of direct methodologies for estimating body fat or
protein. Despite both theoretical and practical problems,
body fat is routinely estimated from a measurement of TBW
using either ^2H_2O or $H_2{}^{18}O$ as tracer and application of
the formula:

$$Body\ fat\ (kg) = Body\ weight\ (kg) - \frac{TBW\ (kg)}{0.732}$$

Inherent in the use of this equation are the assumptions that body weight can be described by the sum of body
fat and lean body mass (fat-free body weight), that body
water resides exclusively in the lean body, and that the
lean body water content is a constant 73.2%[88,89]. One or
more of these assumptions clearly do not hold in mal-
nutrition, obesity, oedema, pregnancy or in trained
athletes, whose body water may account for 72% of <u>total</u>
body weight due to increased muscle mass.

Reliable determination of TBW relies on the rapid
uniform distribution and accurate measurement of the
tracer in a body fluid (plasma, urine or saliva). The
'apparent' dilution space of H_2O is greater than the true
space and greater than that obtained using $H_2{}^{18}O$ as a
result of more general exchange with labile H-atoms of
carboxyl, hydroxyl and amino groups and macromolecules.
The resulting overestimate of TBW will increase lean body
mass estimates as indicated above, and lead to an under-
estimate of body fat. Considerations relating to these
observations have been discussed[90-92]. ^{18}O administered as

$H_2{}^{18}O$ to determine TBW will rapidly equilibrate with the bicarbonate pool of the body (carbonic anhydrase activity) and thus the ^{18}O content of CO_2 in expired air can be used to calculate the water space[65].

Deuterium dilution has recently been employed to quantitate infant breast milk intake by administration of ^2H_2O to the lactating mother and monitoring the appearance of label in the infant[93]. Alternatively, the direct disappearance (dilution) of label can be followed when the tracer is administered to the infant[94].

Theory for the existence of a close relationship between body water and respiratory metabolism and demonstration of the resultant possibility for estimating total CO_2 production (and indirectly energy expenditure) in free-living subjects was first presented some 30 years ago[95]. Briefly, in practical terms, a bolus dose of $^2H_2{}^{18}O$ will rapidly equilibrate with TBW and thence the decreasing concentration of ^{18}O in body water will reflect the loss of both water and carbon dioxide from the body. Simultaneous loss of ^2H will provide a measure of only water loss from the body. The differential disappearance rates of ^2H and ^{18}O provide for the calculation of total CO_2 production and, indirectly, energy expenditure. Validation of the doubly-labelled water method against direct CO_2 production measurement in insects, birds and small animals and man[96,97] has proved very encouraging. Direct applications of the method to such diverse groups as breast fed infants[101] and athletes undergoing heavy sustained exercise[102] have recently been published. Use of the non-invasive $^2H_2{}^{18}O$ method to quantitate energy expenditure in free-living subjects should provide solutions to many apparently paradoxical nutritional 'findings' where published data relating body weight to estimated (by more classical methods) energy intake appear to contravene the laws of thermodynamics.

Substrate cycling

Substrate cycling refers to the simultaneous breakdown and resynthesis of a compound through opposing reactions usually catalysed by different enzymes. These cycles are recognised to play an important part in regulation of carbohydrate and fat metabolism in animals[103,104], but their importance in humans has not been established. Wolfe and colleagues[105] have recently investigated the triglyceride cycle. This cycle releases glycerol and free fatty acids from triglyceride by lipolysis, the triglyceride being reformed by reesterification. By measuring the rate of appearance of glycerol using U-(^{2}H$_5$)glycerol, the rate of appearance of free fatty acids using (^{13}C)palmitate and the rate of fat oxidation by indirect calorimetry, the rates of intracellular, extracellular and total recycling between free fatty acids and glycerol was estimated. Patients with severe burns were thus shown to increase triglyceride cycling by nearly 5 times compared with normal subjects[106]. The energy cost of this recycling contributes to increased thermogenesis and energy requirement in severe burns.

Substrate cycling can also be estimated by constant infusion techniques employing a single, multiply-labelled tracer. We have recently used a constant infusion of (^{2}H$_5$)propionic acid to investigate the reversibility of pathways of propionic acid metabolism. Propionic acid has a number of possible metabolic fates as summarised in Figure 5. The conversion of propionic acid to its various metabolic intermediates will lead to loss of one or more deuterium atoms depending on which intermediate is formed. If propionic acid were to be reformed by reversibility of any of these reactions the hydrogen atoms would almost certainly be taken from body pools at natural abundance, leading to the appearance in plasma of propionic acid of a lower mass than that infused. For example, the formation of acrylyl CoA from propionyl CoA will result in loss of 2 deuterium atoms, so that labelled propionyl CoA recycled via acrylyl CoA will be (^{2}H$_3$) labelled instead of (^{2}H$_5$)labelled. In a normal adult subject, the enrichment of (^{2}H$_1$) to (^{2}H$_4$) propionic acid peaks during the constant infusion of (^{2}H$_5$) propionic acid remained at baseline (Table 3). In a child with methylmalonic aciduria, where the degradation of methylmalonyl CoA to succinyl CoA is impaired due to deficiency of methylmalonyl CoA mutase, an excess enrichment of (^{2}H$_3$)propionic acid was noted during the infusion. This enrichment reached a plateau and was consistent with about 30% of the total propionic acid flux in this child arising from cycling through acrylyl CoA.

Table 3:   Cycling of propionic acid measured using
($^2H_5$)propionic acid

	Plateau enrichment of propionic acid isotopes compared with ($^2H_5$)propionic acid (%)*	
	Normal adult	Child with methylmalonic aciduria
$^2H_1$-propionate	0	0
$^2H_2$-propionate	0	0
$^2H_3$-propionate	0	29
$^2H_4$-propionate	0	0
$^2H_5$-propionate	100	100

* Corrected for isotopic impurity of infusate.   From Walter
JH, Thompson GN, Halliday D, Bartlett K, Leonard JV.
Unpublished data.

Doubly-labelled compounds

By careful choice of label site, it is possible to
use stable isotope labelled compounds to trace the
different metabolic fates of a particular molecule.
Leucine, for example, is known to reversibly transaminate
to its α-ketoacid, α-ketoisocaproic acid (KIC).   This
reaction takes place only intracellularly, principally in
muscle[107].   The further metabolism of KIC ultimately leads
to production of CO_2.   During constant infusion of L-
($^{15}N,1$-^{13}C)leucine, the ^{15}N and ^{13}C-labelled compounds
and the rate of appearance of $^{13}CO_2$ in expired air can be
used to calculate the rates of leucine-KIC transamination
in either direction[108].   By administration of an intravenous bolus of L-($^{15}N,1$-^{13}C)leucine and measurement of
the appearance of (^{13}C)KIC and (^{13}C)leucine in plasma,
similar principles can be used to qualitatively estimate
the rates of transport of leucine and KIC across the cell
membrane – since the (^{13}C)labelled compounds can only be
formed within the cell.   As demonstrated in Figure 7, these
processes appear to proceed extremely rapidly in normal
subjects.

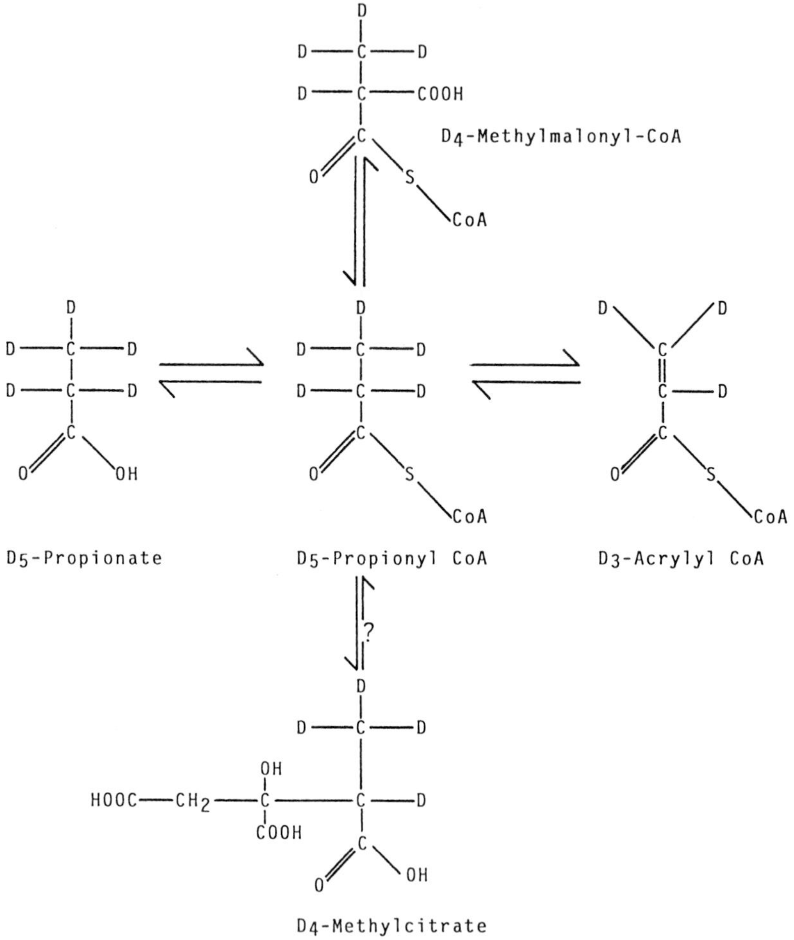

Figure 5: Some possible metabolic interactions of ($^2H_5$) propionic acid. The reversible passage of the propionate skeleton through these reactions will result in differing losses of deuterium label (D).

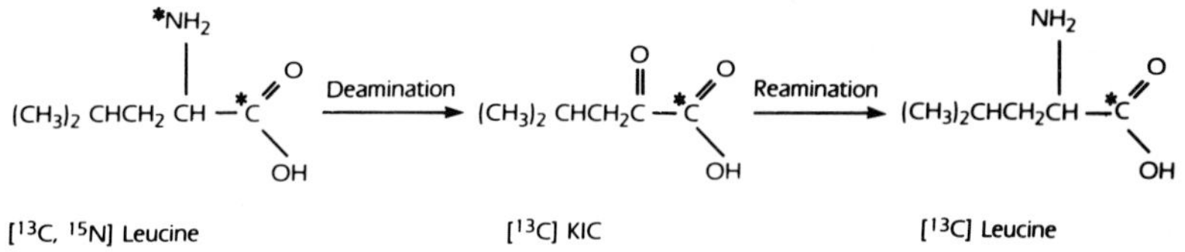

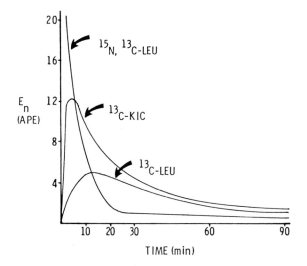

Figure 6: Differing fates of ^{15}N and ^{13}C labels during transamination of leucine. The deamination of L-(^{15}N,1-^{13}C)leucine (M+2) results in loss of the ^{15}N label, so that KIC is singly labelled (M+1). During reamination KIC takes nitrogen from natural abundance body pools, so leucine reappearing is singly labelled (M+1)

Figure 7: Plasma isotopic decay of (^{15}N,1-^{13}C)leucine, ^{13}C-KIC and ^{13}C-leucine after IV bolus of L-(^{15}N,1-^{13}C)leucine. The singly labelled compounds peaked at 6 and 10 minutes respectively. En = enrichment.

Infusion of multiple compounds

Possible limitations of dose of radioactivity do not apply when a number of different stable isotope tracers are to be infused. Furthermore, detection of label in multiple tracees is facilitated by gas chromatography/mass spectrometry. These advantages make possible the comparison of different stable isotope techniques measuring a single process, thereby allowing cross-validation of the techniques. Stable isotopes have been most widely applied in the measurement of protein turnover. A number of methods are available, but the constant infusion of (^{13}C)leucine has become the most favoured because of the short duration (about 4 hours) required. This method, however, requires the measurement of the rate of expired air $^{13}CO_2$ production, necessitating not only isotope ratio mass spectrometry facilities, but also the measurement of total expired air CO_2 production. Clarke and Bier[109] have recently demonstrated that the conversion of phenylalanine to tyrosine can be measured directly by continuous infusions of ($^2H_5$)phenylalanine and (^{13}C)tyrosine. Using a model based on that used for measurement of protein turnover with (^{13}C)leucine, the Clarke and Bier approach can also be applied to the study of protein turnover. Contrary to the findings of Clarke and Bier, it has been demonstrated[110] that plateau enrichments of ($^2H_5$)phenylalanine and of the ($^2H_4$)tyrosine derived from ($^2H_5$)phenylalanine by hydroxylation can be obtained in less than 4 hours by priming of the appropriate pools. Further, ($^2H_2$)tyrosine has been validated as a tracer for tyrosine flux[110]. The combination of these findings has led to the development of a simple and rapid method for measuring protein turnover that requires only the sampling of venous blood, and does not require the infusion of ^{15}N or ^{13}C-labelled tracers. The latter has allowed the comparison with the widely applied (^{13}C)leucine method to be made, and in 6 normal adults, very similar values for whole body protein turnover have been obtained for each technique[110].

Turnover of ketones and carbohydrates

The diverse metabolic fates of glucose present special problems to the measurement of its turnover. Following tissue uptake, glucose can be converted to a number of lower molecular weight compounds such as lactate, which can then be transported back to the liver and recycled to glucose. The measured value for the turnover of glucose in plasma depends on whether the label is recycled. (6,6-$^2H_2$)glucose, for example, will not recycle label through any of the metabolic processes in which glucose normally participates, so this tracer will measure the total flux of glucose including that through tissue recycling, the so called 'true' glucose production rate[111]. If, however, (U-^{13}C)glucose is used as the tracer, recycling will occur; the recycling, however, is not complete due to metabolic 'crossing over' of the label particularly through oxaloacetate. Deuterium labels at the 2- or 3-position in the carbon ring will be lost during other glycolytic-gluconeogenetic cycles[67]. The different fates of various glucose tracers have been used to assess glucose homeostasis in elderly subjects and to show that defective glucose tolerance in these subjects results from a peripheral rather than a hepatic defect[112]. Particular interest has also been generated by this technique in the assessment of glucose metabolism in newborn infants and children. The (6,6-^{2}H$_2$)glucose approach has, for example, been used to assess the value of insulin and glucagon assays, and glucagon and diazoxide administration in hypoglycaemic neonates[113].

The importance of ketone bodies as alternative fuels in the neonatal period has indicated labelled-ketone body turnover studies in this age-group. While earlier studies have been undertaken in adults using radiolabelled tracers, stable isotope techniques, which are ethically indispensable in this age-group, have only recently been developed[114,115]. Using these techniques, Bougneres et al[77] have shown that as much as 25% of neonatal energy requirements may be accounted for by ketone bodies, and that up to 50% of the hepatic ketone body output may be consumed by the central nervous system.

REFERENCES

1. E. White, V.M.J. Welch, T. Sun, L.T. Sniegoski, R. Schaffer,
    H.S. Hertz and A. Cohen, Biomed. Mass Spectrom., 1982, 9, 395.
2. O. Pelletier, L.A. Wright and W.C. Breckenridge, Clin. Chem.,
    1987, 33, 1403.
3. H.S. Taylor, W.W. Swingle, H. Eyring et al., J. Chem. Phys.,
    1933, 1, 751.
4. G.N. Lewis, Science, 1934, 79, 151.
5. J.F. Thompson, 'Biological effects of deuterium', Macmillan,
    New York 1963, p 133.
6. C.T. Gregg, J.Y. Hutson, J.R. Prime et al., Life Sci., 1973
    13, 775.
7. D. Wolf, H. Cohen, A. Meshorer et al., 'Stable isotopes'.
    Proceedings of the Third International Conference. E.R. Klein,
    P.D. Klein (eds): Academic Press, New York, 1979, p 360.
8. D.L. Smith, Anal. Chem., 1983, 55, 2391.
9. J. Eagles, S.J. Fairweather-Tait and R. Self, Anal. Chem., 1985
    57, 469.
10. A.L. Gray and A.R. Date, Analyst, 1983, 108, 1033.
11. D.E. Matthews and J.M. Hayes, Anal. Chem., 1978, 50, 1465.
12. J.W. Gramlich, L.A. Machlan, K.A. Brletic and W.R. Kelly, Clin.
    Chem., 1982, 28, 1309.
13. M. Janghorbani, V.R. Young, J.W. Gramlich and L.A. Machlan,
    Clin. Chem. Acta , 1981, 114, 163.
14. L.J. Moore, L.A. Machlan, M.O. Lim, A.L. Yergey and J.W. Hansen,
    Paediatr. Res., 1985, 19, 329.
15. J.R. Turnlund, J.C. King, W.R. Keyes, B. Gong and M.C. Michel,
    Am. J. Clin. Nutr., 1984, 40, 1071.
16. M.B. Rabinowitz, G.W. Wetherill and J.D. Kopple, Science, 1973,
    182, 725.
17. J. Graff and D. Rittenberg, Anal. Chem., 1952, 24, 878.
18. M.L. Coleman, T.J. Shepard, J.J. Durham, J.E. Rouse and G.R.
    Moore, Anal. Chem., 1982, 54, 995.
19. D. Fraisse, D. Girard and R. Levy, Talanta, 1973, 20, 667.
20. I. Friedman, Geochim. Cosmochim. Acta, 1953, 4, 89.
21. D. Halliday and A.G. Miller, Biomed. Mass Spectrom., 1977, 4, 82.
22. C.C. Harris, D.M. Smith and J. Mitchell, Anal. Chem., 1950, 22,
    1297.
23. K.G. Hardcastle and I. Friedman, Geophys. Res. Lett., 1974, 1,
    165.
24. T.W. Boulton, W.W. Wong, D.L. Hackey, L.S. Lee, M.P. Cabrera and
    P.D. Klein, Anal. Chem., 1983, 55, 1832.
25. B.N. Smith and S. Epstein, Plant Physiol., 1970, 46, 738.
26. W.W Wong and P.D. Klein, Mass Spec. Rev., 1986, 5, 313.
27. M. Sano, Y. Yotsui, H. Abe and S. Sasaki, Biomed. Mass Spectrom.,
    1976, 3, 1.
28. A. Otsuki, Y. Ino and T. Fuju, Int. J. Mass Spec. Ion Phys..
    48, 343.

29.   A. Barrie, J. Bricout and J. Koziet, Biomed. Mass Spectrom.,
      1984, 11, 583.
30.   A.M. Ajami and J.B. Watkins, Clin. Chem., 1983, 29, 725.
31.   D.A. Schoeller, P.D. Klein and J.F. Schneider, Proceedings of
      the Second International Conference on Stable Isotopes.
      E.R. Klein, P.D. Klein (eds): Academic Press, New York, 1975,
      p 246.
32.   P.J. Ross and A.E. Martin, Analyst, 1970, 95, 817.
33.   L.K. Porter and W.A. O'Deen, Anal. Chem., 1977, 49, 514.
34.   W.W.C. Read, R.A. Harrison and D. Halliday, Anal. Biochem., 1982,
      123, 249.
35.   O. Teuji, M. Masugi and Y. Kosai, Anal.Biochem., 1975, 65, 19.
36.   T. Preston and J.P. Owens, Analyst, 1983, 108, 971.
37.   A. Barrie and C.T. Workman, Spectros. Int. J., 1984, 3, 439.
38.   W.W. Wong, M.P. Cabrera and P.D. Klein, Anal. Chem., 1984, 56
      1852.
39.   A. Barrie and W.A. Coward, Biomed. Mass Spectrom., 1985, 12, 535.
40.   S. Epstein and T. Mayeda, Geochim Cosmochim. Acta, 1953, 14, 213.
41.   W. W. Wong, L.S. Lee and P.D. Klein, Anal. Chem., 1987, 59, 690.
42.   H. Puchett, B.R. Sabels and T.C. Hoering, Cosmochim. Acta, 1971,
      35, 625.   ︀
43.   W.F. Schwenk, P.J. Berg, B. Beaufrere, J.M. Miles and M.W.
      Haymond, Anal. Biochem. 1984, 141, 101.
44.   R.A. Shipley and R.E. Clark. 'Tracer methods for in vivo
      kinetics', Academic Press, New York, 1972, p 1.
45.   D.J. Millward, B. De Benoist and D. Halliday, Clin. Nutr. Met.
      Res., 'Proceedings 7th Congress ESPEN Dietze, Grunert Klinberger
      and Wolfram (eds): Karger, Basel, 1986, p 178.
46.   V.R. Young, Am.J. Clin. Nutr., 1987, 46, 709.
47.   D. Halliday and R.O. McKeran, Clin. Sci. Mol. Med., 1975, 49, 581.
48.   J.M. Conway, D.M. Bier, K. Motil, M. Burke and V.R. Young, Am.J.
      Physiol., 1980, 239, E192.
49.   M. Gersovitz, D. Bier, D. Matthews, J. Udall, H.N. Munro and
      V.R. Young, Metabolism, 1980, 29, 1087.
50.   D.E. Matthews, J.M. Conway, V.R. Young and D.M. Bier, Metabolism,
      1981, 30, 886.
51.   M.A. Staten, D.M. Bier and D.E. Matthews, Am. J. Clin. Nutr.,
      1984, 40, 1124.
52.   M.M. Meguid, D.E. Matthews, D.M. Bier, C.N. Meredith and V.R.
      Young, Am. J. Clin. Nutr., 1986, 43, 781.
53.   X-L.Zhao, Z-M. Wen, C.M. Meredith, D.E. Matthews, D.M. Bier and
      V.R. Young, Am. J. Clin. Nutr., 1986, 43, 795.
54.   D.E. Matthews, K.J. Motil, D.K. Rohrbaugh, J.F. Burke, V.R. Young
      and D.M. Bier, Am. J. Physiol., 1980, 238, E473.
55.   M.M. Meguid, D.E. Matthews. D.M. Bier, C.N. Meredith, J.S. Soeldner
      and V.R. Young, Am. J. Clin. Nutr., 1986, 43, 770.
56.   V.R. Young, C. Gucalp, W.M. Rand, D.E. Matthews and D.M. Bier,
      Hum. Nutr: Clin. Nutr., 1987, 41C, 1.

57. R.D. Yang, D.E. Matthews, D.M. Bier and V.R. Young, Am. J.
     Physiol., 1984, 247, E634.
58. M. Gersovitz, H.N. Munro, J. Udall and V.R. Young, Metabolism,
     1980, 29, 1075.
59. T.P. Stein, J.L. Mullen, J.C. Oram-Smith, E.F. Rosato, H.W.
     Wallace and W.C. Hargrove, Am. J. Physiol., 1978, 234, E648.
60. M.J. Rennie, R.H.T. Edwards, D.J. Millward, S.L. Wolman,
     D. Halliday and D.E. Matthews, Nature, 1982, 296, 165.
61. J.N.A. Gibson, D. Halliday, W.L. Morrison, P.J. Stoward, G.A.
     Hornsby, P.W. Watt, G. Murdoch and M.J. Rennie, Clin. Sci., 1987,
     72, 503.
62. K.N. Cheng, F. Dworzak, G.C. Ford, M.J. Rennie and D. Halliday,
     Eur. J. Clin. Invest., 1985, 15, 349.
63. J.C. Waterlow, M.H.N. Golden and P.J. Garlick, Am. J. Physiol.,
     1978, 235, E165.
64. D.A. Schoeller, E. van Santen, D.W. Peterson, W. Dietz,
     J. Jaspan and P.D. Klein, Am. J. Clin. Nutr., 1980, 33, 2686.
65. D.A. Schoeller and E. van Santen, Am. J. Physiol., 1982, 53,
     955.
66. F.K. Trefz, T. Erlenmaier, D.H. Hunneman, K. Bartholomé and
     P. Lutz, Clin. Chim. Acta, 1979, 99, 211.
67. R.R. Wolfe, 'Tracers in metabolic research: radioisotope and
     stable isotope/mass spectrometry methods'. Alan R. Liss,
     New York, 1984.
68. J.S. Garrow and J.D. Webster, Hum. Nutr: Clin. Nutr., 1986, 40C,
     315.
69. P.J. Garlick, M.A. McNurlan, K.C. McHardy, A.G. Calder, E. Milne,
     L.M. Fearns and J. Broom, Hum. Nutr: Clin. Nutr., 1987, 41C,
     177.
70. J. Katz, J. Fed. Proc., 1982, 41, 123.
71. G.N. Thompson, P.J. Pacy, G.C. Ford, H. Merritt and D. Halliday,
     Unpublished observations, 1988.
72. R.R. Wolfe, J.E. Evans, C.J. Mullany, J.F. Burke, Biomed. Mass
     Spectrom., 1980, 7 168.
73. F. Pirney, M. Lecroix, F. Mosora, A. Luyckx and P. Lefebure,
     J. App. Physiol., 1977, 43, 258.
74. E. Ravussin, P. Pahud, A. Thelin-Doerner, M.J. Arnaud and E.
     Jequier, Int. J. Obesity, 1980, 4, 235.
75. D.A. Schoeller, J.F. Schneider, N.W. Solomons, J.B. Watkins and
     P.D. Klein, J. Lab. Clin. Med., 1977, 90, 412.
76. D.A. Schoeller and P.D. Klein, Biomed. Mass Spectrom., 1978, 6,
     350.
77. P.F. Bougneres, C. Lemmel and D.M. Bier, J. Clin. Invest., 1986,
     77, 42.
78. D.A. Schoeller, C. Brown, K. Nakamura, A. Nakagawa, R.S. Mazzeo,
     G.A. Brooks and T.F. Budinger, Biomed. Mass Spec., 1984, 11, 557.
79. N.W. Solomons, D.A. Schoeller, J.B. Wagonfeld, D. Ott, I.H.
     Romberg and P.D. Klein, J. Lab. Clin. Med., 1977, 90, 431.

80. B.H. Lauterberg and J. Bircher, J. Pharmacol. Exp. Therapy., 1976, 196, 501.

81. W.W. Shreeve, J.D. Shoop, D.G. Ott and B.B. McInteer, Gastroenterology, 1976, 71, 98.

82. J.B. Watkins, D.A. Schoeller, P.D. Klein, D.G. Ott, A.D. Newcomer and A.F. Hofman, J. Lab. Clin. Med., 1977, 90, 422.

83. D.Y. Graham, P.D. Klein, D.J. Evans, D.G. Evans, L.C. Alpert, A.R. Opekun and T.W. Boutton, Lancet, 1987, 1174.

84. E.A. Bering, J. Neurosurg., 1952, 9, 275.

85. W.A. Neely, M.D. Turner, J.D. Smith and J. Williams, Surg. Forum, 1961, 12, 13.

86. J.F. Scholer and C.F. Code, Gastroenterology, 1954, 27, 565.

87. H-L. Schmidt, P. Kirch, G. Traul and H-E. Keller, 'Proceedings of the Second International Conference on Stable Isotopes', E.R. Klein, P.D. Klein (eds): Academic Press, New York, 1975, p 411.

88. N. Pace and E.N. Rathbun, J. Biol. Chem., 1945, 158, 685.

89. F.B. Forbes, 'Human Growth', vol. 2, F. Falkner and J.M. Tanner (eds): Plenum, New York, 1978, p 239.

90. J.M. Culebras and F.O. Moore, A. J. Physiol., 1977, 232, R54.

91. J.M. Culebras, G.F. Fitzpatrick, M.F. Brennan, C.M. Boyden and F.O. Moore, Am. J. Physiol., 1977, 232, R60.

92. R.K. Whyte, H.S. Bayley and H.P. Schwarcz, Am. J. Clin. Nutr., 1985, 41, 801.

93. W.A. Coward, T.J. Cole, M.B. Sawyer and A.M. Prentice, Hum. Nutr: Clin. Nutr., 1982, 36C, 141.

94. W.A. Coward, M.B. Sawyer, R.G. Whitehead, A.M. Prentice and J. Evans, Lancet, 1979, 13.

95. N. Lifson, G.B. Gordon and R. McClintock, J. App. Physiol., 1955, 7, 704.

96. J.B. Williams and K.A. Nagy, Physiol. Zool., 1984, 57, 325.

97. K.A. Nagy, Am. J. Physiol., 1980, 238, R466.

98. D.A. Schoeller and E. van Staten, J. Appl. Physiol., 1982, 53, 955.

99. P.D. Klein, W.P.T. James, W.W. Wong, C.S. Irving, P.R. Murgatroyd, M. Cabrera, H.M. Dallosso, E.R. Klein and B.L. Nichols, Hum. Nutr: Clin. Nutr., 1984, 38C, 95.

100. S.B. Roberts, W.A. Coward, K.-H. Schlingenseipen, V. Nohria and A. Lucas, Am. J. Clin. Nutr., 1986, 44, 315.

101. A. Lucas, G. Ewing, S.B. Roberts and W.A. Coward, Brit. Med. J., 1987, 295, 75.

102. K.R. Westerterp, W.H.M. Saris, M. van Es and F. ten Hoor, J. Appl. Physiol., 1986, 61, 2162.

103. B. J. Brooks, J.R.S. Arch and E.A. Newsholme, Biosci. Rep., 1983, 3, 263-7.

104. B. Issekutz, Metabolism, 1977, 26, 157-70.

105. R.R. Wolfe and E.J. Peters, Am. J. Physiol., 1987, E218-23.

106. R.R. Wolfe, D.N. Herndon, F. Jahoor, H. Miyoshi and M. Wolfe, New Eng. J. Med., 1987, 317, 403.

107. A.E. Harper and C. Zapalowski, in ' Metabolism and Clinical Implications of Branched Chain Amino and Ketoacids', M. Walser, J.R. Williamson (eds): Elsevier North Holland, New York, 1981, p 195.

108. D.E. Matthews, D.M. Bier, M.J. Rennie, D. Halliday, R.H.T. Edwards, D.J. Millward and G.A. Clugston, Science, 1981, 214, 1129.

109. J.T.R. Clarke and D.M. Bier, Metabolism, 1982, 31, 999.

110. G.N. Thompson, P.J. Pacy, G.C. Ford, H. Merritt and D. Halliday, 1987, Unpublished observations.

111. D.M. Bier, R.D. Leake, M.W. Haymond, K.J. Arnold, L.D. Gruenke, M.A. Sperling and D.M. Kipnis, Diabetes, 1977, 26, 1016.

112. J.J. Robert, J.C. Cummins, R.R. Wolfe, M. Durkot, D. Matthews, X-H Zhao, D.M. Bier and V.R. Young, Diabetes, 1982, 31, 203.

113. A. Mehta, R. Wootton, K.N. Cheng, P. Penfold, D. Halliday and T.E. Stacey, Arch. Dis. Childhood, 1987, 62, 924.

114. J.M. Miles, W.F. Schwenk, K.L. McLean and M.W. Haymond, Am. J. Physiol., 1986, 251, E185.

115. P.F. Bougneres, E.O. Balasse, P. Ferre and D.M. Bier, J. Lipid Res., 1986, 27, 215.

8
Radiopharmaceuticals

K. Kristensen

THE ISOTOPE-PHARMACY, THE NATIONAL BOARD OF HEALTH, 2700 BRØNSHØI, DENMARK

1 INTRODUCTION

Radiopharmaceuticals are radioactive material intended for use in medical diagnosis or therapy. They are a group of pharmaceuticals comprising of a huge variety of compounds from simple carrier-free inorganic salt solutions to radioactive labelled proteins such as antibodies prepared by monoclonal technique. The main use is in medical diagnosis. The radionuclide in the compound makes the use as a tracer substance possible. The radiation to the human body is an inescapable but unwanted side effect. In the case of radiopharmaceuticals for therapy, however, the radiation effect is the effect wanted.

There are two main types of radiation emission used in nuclear medicine. Radionuclides that emit single gamma-photons (often in combination with electrons). For in vivo measurements this is the absolute dominating type of radiopharmaceuticals. For the detection a gamma-camera or single crystal detectors all based on scintillation detectors are used. Special equipped gammacameras may also be used for single photon emission computerized tomography (SPECT). The second type is radionuclides that emit positrons which in turn are converted into two gamma rays. This type of radionuclides are used for positron emission tomography.

The following description of radiopharmaceuticals and of current developments treat these two groups separately. Radiopharmaceuticals for therapy are also

treated separately. It is not the intention in this
review to cover all available radiopharmaceuticals but
mainly to illustrate lines of development.
The classification of radiopharmaceuticals in this review is not necessarily logical as different types of
classification are used simultaneously. It has' been
done in this way to be able better to describe lines of
development. This survey describes therefore some of
the most important radiopharmaceuticals but it is in no
way complete.

2 RADIOPHARMACEUTICALS FOR DIAGNOSTIC USE (SINGLE PHOTON EMISSION)

2.1 Simple salt-solutions

Table 1   gives a list of the most important radiopharmaceuticals used in the form of simple salt
solutions. The basis for the use may be their natural
occurrence in the human body. This applies to radionuclides of elements   such as sodium, potassium,
calcium, iron and iodine. Several of these radionuclides do not have gamma energy characteristics that
fit with modern nuclear medicine instrumentation.
Analogues are therefore wanted. Thallium is used in
heart studies based on its biological behaviour (1)
similar to potassium and the binding of Indium to
transferrin is similar to that of iron.
However, the main importance of radionuclides of
chromium, indium and iodine comes from their ability to
be used as labelling agent for biological material and
other organic molecules.
 Two radionuclides: Gallium and Technetium are used
because of special biological behaviour and good energy
characteristics. Gallium-67 is accumulated in several
tumors and in inflammatory processes. There are many
theories about the mechanism and several factors such
as perfusion, diffusion, permeability and binding to
cell elements seem important.(2)
Technetium-99m is the most important radionuclide in
modern nuclear medicine. In most countries it is used
in more than fifty percent of all procedures carried
out. Its monoenergetic radiation of 140 keV is well
suited for the gammacamera.In the form of pertechnetate
it has been used in brain tumor detection. The
mechanism is not completely known but it is based on
the difference in permeability in normal brain and
other tissues. The pertechnetate ion is of similar

ionic radius as the iodide ion and both are bound to the same tissues such as thyroid, salivary glands and gastric mucosa. It can be used as a replacement for iodide ion thyroid diagnostics although only the iodide trapping mechanism can be measured as pertechnetate is not metabolized as iodide.(3).

Table 1   Radiopharmaceuticals in the form of simple salt solutions

Radio-nuclide	Chemical form	$T_{1/2}$	Principal gamma energy	Production
Na-22	Sodium chloride	2.6y	511 keV	C
Na-24	-	15 h	1369 -	R
P-32	Sodium phosphate	14.3d	-	R
K-42	Potassium chloride	12.3h	1524 -	R
Ca-47	Calcium chloride	4.5d	1308 -	R
Cr-51	Sodium chromate	28 d	511 -	R
	Chromic chloride	28 d	511 -	R
Fe-59	Ferrous citrate	45 d	1095 -	R
Ga-67	Gallium citrate	3.2d	93 -	C
Tc-99m	Sodium pertechnetate	6 h	141 -	G
In-111	Indium chloride	2.8d	247 -	C
	Indium citrate	2.8d	247 -	C
In-113m	Indium chloride	1.6h	393 -	G
I-123	Sodium iodide	13.2h	159 -	C
I-125	-	60 d	35 -	R
I-131	-	8.1d	364 -	R
Tl-201	Thallous chloride	3.1d	167 -	C

C : cyclotron,   R : reactor,   G : generator system
Apart from the pioneering work of Hevesy and others with phosphorous radionuclides it was the availability of Iodine-131 in large amounts that made the real nuclear medicine techniques possible. It is still one of the cornerstones even if the high gamma energy emitted is not very suitable for modern instrumentation. It is, however, one of the few that at the same time can be used in therapy due to its emission of beta-particles.
There are more than 20 radionuclides of iodine, but only a few are useful in nuclear medicine. Iodine-125 emits only gamma-radiation too weak to be useful for in vivo measurement. Newest development is Iodine-123. The twelve hour half life gives logistic problems but this radionuclide has a very good energy for use with gammacameras. The saving in radiation dose compared to

I-131 is also considerable. Different production methods lead to products of different radionuclidic purity. I-124 and I-125 are main impurities in I-123. With the use of Xe-124 as a target it is now possible to produce I-123 with a radionuclidic purity of more than 99.99%. This requires 100% enriched Xe-124 as a target for irradiation. Natural xenon contains 0.096% Xe-124 (4).

Although sodium radioiodide still is used for thyroid function studies the main use of iodine radionuclides is for labelling of organic compounds and material of biological origin.

The possibilities for developing new radionuclides are of course very limited but there are still possibilities for the industry to make available radionuclides such as I-123 which only has been generally available for a few years. Particularly there may be possibilities for the development of generator systems based on mother/daughter decay schemes.

2.2 Radioactive gases

Xenon-133 is a non-metabolized noble gas used either as a gas for ventilation studies of the lungs or dissolved in isotonic saline for perfusion studies of different organs including blood flow studies of the brain. The gamma energy is rather low and Xenon-127 is to be preferred. It is, however, more expensive and should preferably be re-used. Xenon-133 is readily available in large quantities in connection with the production by a fission process of molybdenum-99 for Tc-99m generators.

Table 2 Radioactive, chemically inert gases.

Radionuclide	Half-life	Principal gamma-energy
Krypton-81m	13 s	190 KeV
Xenon-127	36,4 d	203 -
Xenon-133	5,3 h	81 -

The most recent development is Krypton-81m. It is a daughter of Rubidium ($T_{1/2}$ 4.6 hours) and must be prepared direct from a generator either for inhalation or for infusion. In the last mentioned form it may be used for cardiac function studies. Direct infusion generators such as the Rb81/Kr81m generator where the generator eluate is injected directly into the patient do of course present quality control problems (5).

2.3 Radiopharmaceuticals as colloids, particles and aerosols

A particulate tracer injected intravenously will first pass the heart and be mixed completely with blood and thereafter distributed throughout the lung in proportion to blood flow. If the tracer consists of particles too large to pass the lung capillaries they will be trapped during their first passage. This is the basis for the use of macroaggregated human serum albumin and microspheres for lung perfusion studies. Smaller particles and colloids will pass through the lungs and will be taken up by the reticuloendothelial system in the liver spleen and bone marrow. The relative distribution depends on the blood flow to the various organs. The size of the colloids may effect their distribution. Smaller particles have a greater accumulation in the marrow while larger particles may accumulate to a relatively higher degree in the spleen.

Table 3 Tc-99m Radiopharmaceuticals based on physical /chemical principles for biodistribution.

Type	Preparation	Range of mean particle size
Colloid	Sulphur colloid	300-500 mm
	Tin colloid	150-250 mm
	Rhenium sulfide-colloid	40 mm
	Antimony sulfide-colloid	3-30 mm
	Albumin colloid	5-100 mm
Particles		
	Macroaggregated albumin	10-50 p.m.
	Microspheres	10-20 p.m.
Aerosoles		
	Nebulized DTPA or. sulphur colloid	
	Pseudogas -	

Macroaggregates - Microspheres. The ideal particle size for lung perfusion studies would be 10-20 μm. Both preparations are made from human serum albumin and are therefore bio-degradable. Microspheres are made by heating an oil emulsion. They may be separated in rather narrow particle ranges by sieving. Macro-aggregates are denatured by heat and pH adjustment. The particle size range is larger but they are cheaper to prepare and for routine clinical use they are of equal

value. Both are available in kit form for labelling
with Tc-99m. The number of particles injected may
affect the results (and the safety). A sufficient
number of particles (appr. 10^5) to give an even
distribution of radioactivity should be given. On the
other hand one should not block more lung capillaries
than necessary.(5)

Aerosols. Particles used for perfusion scintigraphy or
a solution of a Tc-99m labelled radiopharmaceutical
such as DTPA or sulphur colloid may be made into an
aerosol by a nebulizer and used as a replacement for a
radioactive gas in lung ventilation studies. It is a
rather complicated procedure seen from a radiation
hygiene and quality control point of view. The success
in use depends very much on the reproducibility of the
procedure used. A recent development is the
Technegas (6). A Tc-99m-solution is in a special
furnace heated to form particles (with carbon ?) and it
can then be used as an alternative to other aerosols or
radioactive gases.

Colloids. There exist a number of different labelling
methods for the preparation of Tc-99m colloids which
are the only colloidal radiopharmaceuticals of real
importance. These colloids have very different
particle size and probably different surface
properties. Sulphur colloids, rhenium and antimony
sulphide colloids are labelled as they are prepared
during a heating process while tin colloid and albumin
microcolloids are preformed and labelled in a one step
kit form. For the visualisation of liver and spleen the
particle size is not critical and quantitative studies
with colloids are not much used. For use in
lymphoscintigraphy particle size becomes much more
important. Particles smaller than a few nanometer may
pass into the systemic circulation while particles
larger than 15 nanometer are cleared slowly from the
injection site. Tc-99m antimony sulphide colloids seem
to be most useful.(8)

2.4 Radiopharmaceuticals from human materials

In the line of developments of radiopharmaceuticals the
labelling of human materials belongs to one of the
earlier phases but also to one of the latest. The
iodination of albumin was a pioneer work and gave a lot
of experience that can be used when studying the effect
of introducing a foreign label in a large molecule.
Among the latest development is the labelling of cells
with a I-123 labelled antibody.

Table 4 gives examples of human substances labelled and of the radionuclide used.

Table 4 Radioactive labelled human material.

Material labelled	Labelling material
Albumin	Cr-51:chloride,chromate
Fibrinogen	Tc-99m:pertechnetate/
Cells: Red blood cells	tin chloride, HMPAO
leucocytes, Platelets	pyrophosphate, MDP
	In-111: oxine,
	tropolone, DTPA
	I-125/I-131: iodide/
	chloramine-T, iodogen

In all cases the label is a foreign atom introduced into the molecule. The behaviour of the reintroduced labelled molecule is therefore not necessarily the same as the native molecule, but the experience is that it is possible to prepare useful material. The most useful radiopharmaceuticals recently prepared in this way are red blood cells labelled with Tc-99m for cardiac studies(8) and In-oxine or Tc-99m HMPAO labelled white cells for detection of infection sites.(9).
It is possible to label human serum albumin with I-131 in such a way that near "metabolic" substances can be obtained when the iodine content is restricted to one atom per molecule. Tc-99m labelled human serum albumin preparations available have been shown to be very different in properties and to consist of different fractions with different biological properties (10).

2.5 Radiopharmaceuticals: Excretory function
The two main areas are the kidneys and the hepatobiliary system.
Kidneys. Table 5 lists some of the radiopharmaceuticals in current use. I-131-ortoiodohippuric acid has for a long time been considered the most relevant substance for studying kidney function.(11). I-131 is not very suitable for use with the gammacamera. A search for a Tc-99m labelled analogues has been extensive. An alternative now is I-123 ortoiodohippuric acid where a more suitable radionuclide is included but with the drawback of the 12 hour half-life. Glomerular filtration studies can be carried out with Tc-99m DTPA. However, it has an extraction efficiency of only 20% and it is therefore not well ˈsuited for studying kidneys with very low function.

Table 5   Radiopharmaceuticals   for   kidney   function
studies

Cr-51	EDTA complex
Tc-99m	DTPA complex
	Glucoheptonate
	MAG$_3$
I-123	Sodium iodohippurate
I-131	Sodium iodohippurate

A new class of chelating agents has been introduced
based on DADS, a diamide disulphur ligand. A number of
different analogues were Tc-99m labelled and provided
information on structure/activity relationship. Tc-99m
CO$_2$DADS forms two isomers on labelling and needs
separation before use. It is therefore not clinically
useful (12). A new compound MAG$_3$ (mercaptoacetyltri-
glycerine) avoids this problem and may become a useful
renal agent (13,14).

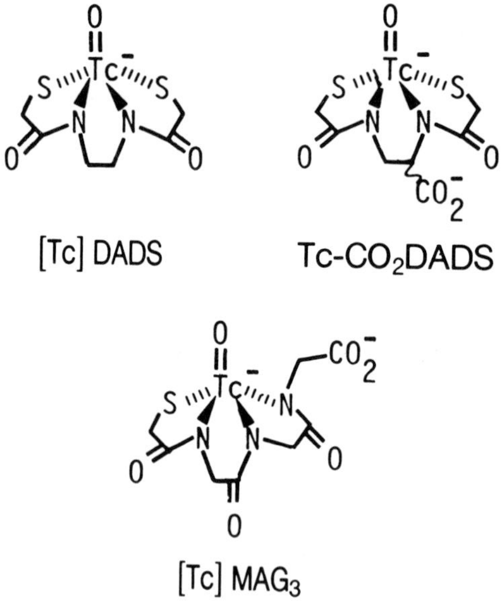

[Tc] DADS   Tc-CO$_2$DADS

[Tc] MAG$_3$

Figure 1: Structures of Tc-N$_2$S$_2$ (DADS) complexes and
           proposed structure of Tc-MAG$_3$ (13).
           Reproduced with permission of the Journal of
           Nuclear Medicine.

Hepatobiliary system. A series of phenylsubstituted derivatives has been developed and in Table 6 is listed a number of available preparation kits for Tc-99m labelling. This research provided systematic information on the effect of substition in the molecule and its relation to heptobiliary excretion. The information was, however, often of a mixed nature and illustrates the difficulties structure/biodistribution considerations must face. The information, however, now is piling up and in general renal excretion of Tc-99m complexes decreases with increasing lipophilicity and hepatic excretion increases. It is now from this type of research possible to forsee the type of combinations of substitutions that might lead to a more optimal hepatobiliary agent. The resulting compound Tc-99m trimethylbromoiminodiacetic acid was indeed superior with regard to specificity and pharmacokinetic parameters (15,16).

Table 6 List of derivatives for hepatobiliary function test available as Tc-99m labelling kits.

Dimethyl	–	acetanilidoiminodiacetate
Diethyl	–	–
Diisopropyl	–	–
Diethyliodo	–	–
Trimethyl	–	–

2.6 Radiopharmaceuticals: Cell/organ uptake.

Brain: Advances in Positron emission tomography (PET) are being translated into single photon emission tomography (SPECT) particularly in the brain imaging area, where also another starting point has been the blood flow measurements with radioactive gases (Xe-133). This transfer of techniques takes place with I-123 compounds such as P-iodoamphetamine and hydroxy trimethyliodobenzylpropandiamine (HIPDM). Tl-201-diethyldithiocarbamate (DDC) is another approach but the most successful so far has been the Tc-99m labelled compounds that is a result of a long and systematic search for a lipophilic compound that would cross the blood/brain barrier and stay in the brain sufficiently long for measurement with the gammacamera type of instrumentation (17).
Tc-99m propyleneamineoxime (PnAO) is a neutral lipophilic complex with such properties; however the washout is rather fast. A whole range of substituted compounds on this backbone structure was developed by Amersham Int. Hexamethylpropyleneamine oxime labelled compound showed stereoisomerism and there were large

differences in the biodistribution. The d,l form showed the highest uptake.(18)   The complex is a Tc(V)oxo complex and it is _in vitro_ and _in vivo_ transformed into a hydrophilic complex that is unable to leave the brain again. It seems that this conversion is a reaction with glutathione.(19)   It takes also place in blood.

Tc-HM-PAO

Fig.2 Proposed structure of the lipophilic Tc-HM-PAO
     and the hydrophilic secondary complex (20).
   The labelling with Sn^{2+} must be carried out under standardised conditions and the Tc-99m pertechnetate solution must be freshly prepared. The labelled complex must be used within 30 minutes as the conversion also takes place _in vitro_.
Another group selected the N_2S_2-structure   and a number of Tc(V)oxo complexes has been prepared. N-piperidinyl-DADT complexes showed the highest uptake in brain. The preparations developed   so   far   require   however purification by HPLC before use and it is cleared relatively fast from the brain (21). It is possible that the development of small neutral complexes with an appropriate side group attached to promote binding to cellular counterparts may pave the way for specific receptor binding Tc-99m radiotracers.
Heart: A Tc-99m replacement for Tl-201 for myocardial perfusion has been looked for. A whole class of Tc-99m phosphine complexes (DMPE) has been developed with technetium in I, III and V oxidation state. This research was the first to use the low specific activity Tc-99 for the chemical structure characterization of compounds and the transferal of this knowledge to the Tc-99m labelled compounds by HPLC comparison. HPLC being able to cope with both types of products. It also illustrated the species differences between different animals and between animals and humans. Several compounds useful in animals did have completely different biodistribution in humans.(20).

More successful was the screening of Tc-99m isonitrile derivatives. Tc-99m hexakis-2-methoxy-2-methyl propyl-1-isonitrile (RP30A). It has shown favourable myocardial uptake and blood clearance in animals and humans. Again a group of derivatives are being prepared and studied to increase our knowledge of chemical structure/biodistribution.(22).

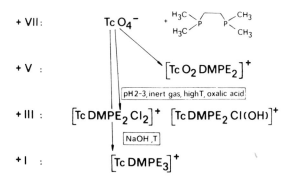

Fig. 3. Reaction routes for the system TcO$_4$/DMPE (Reproduced from ref.20 with permission from Nijhof Publishers.

Bone. The development of Tc-99m compounds in this area started with the discovery of the binding properties of polyphosphates for reduced technetium. It was shown that the long chain polyphosphate could be replaced by pyrophosphate or its analogues: Imidophosphonate or diphosphonate.(15) Methylendiphosphonate has so far been selected as the most clinical useful compound with highest bone uptake and lowest blood background. The preparation kits available give a mixture of different complexes. Pure Tc-99m complexes have been characterized but they have not been developed into commercially available products.

2.7 Radiopharmaceuticals: Metabolic function

I-131-m-iodobenzylguanidine. This guanithidene analogue is capable of localizing in the adrenal medulla. It has structure similar to adrenalin. It has been shown to be useful as a means of identifying malignant phaeochromocytoma. An example of a rather specific radiopharmaceutical.(23).

Cholesterol derivatives. Iodinated (I-131) cholesterol
derivatives have been used for adrenal scintigraphy for
a number of years.(24). An alternative with better
radiation characteristics, better biological clearance
and better in vitro stability has been developed in the
form of 6-B-methylselenomethyl-19-norcholesterol-5(10)-
en 3 B-ol-Se75. An example that other foreign radionuclides than iodine may be introduced in large organic
molecules without changing the biodistribution
dramatically (25).

Fatty acids. Fatty acids are particularly metabolised
by the myocardium I-131 and lately I-123 labelled fatty
acid has therefore been developed for imaging in myocardial infarction (26).
A long list of compounds are available and
structure/metabolism has been studied. Particularly I-
123 heptadecanoic acid and I-123 pentadecanoic acid
have been studied clinically.

2.8 Receptor based radiopharmaceuticals

Receptors have been described as high binding affinity
proteins with a specific anatomical distribution that
responds to low levels of specific substances with a
defined physiological event. Receptor based radiopharmaceuticals are potential tools for the study of
physiology and biochemistry in vivo.(27) Changes in
the interaction of a ligand with its receptor may have
implications in human disease. Estrogen receptors have
been imaged in vivo by the use of F-18 estradiol.
Dopamine receptors were identified in pituitary tumors
by a carbon 11 labelled compound. I-123 labelled
methylspiperone for the study of dopamine receptors is
a third example of a recent development which brings
this area out of the complicated and economically
restricted area of cyclotrons and positron emission
tomographs.(PET).   A main problem is to obtain radiopharmaceuticals of sufficient high specific activity.
The concentration of receptors is very low and often in
the order of picomole per ml. The radioactive
concentration needed for imaging may be 1-10 kBq/cc. It
is further anticipated that not more than 10% of
receptors can be occupied before a pharmacological
response may be seen. Specific activities in   excess
of 100 Giga Bq/mmol may therefore be needed. The very
short half-life of the relevant radionuclides present
the chemist with a special challenge and special
synthetic methods must be developed. High performance
liquid chromatography is particularly useful for
purification and characterization of such products.

2.9   Monoclonal antibodies

The technique for producing large amounts of highly specific monoclonal antibodies has been developed during the last ten years. In principle this technique has unlimited potential also in nuclear medicine for diagnostic and therapeutic use.

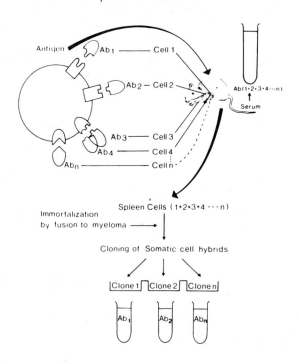

Fig.4 Principles of the production of monoclonal anti-
       bodies by hybridoma technique (28).
       (Reproduced   with   permission   from   Nijhof
       Publishers).

The production involves in principle the following steps: A mouse is injected with the antigen and it produces B lymphocytes that can form antibodies. The spleen of this mouse rich in B lymphocytes is removed and mixed with mouse myeloma cells. A fusing cell such as polythyteneglycol is added in a medium where only the hybrids will survive. The hybridoma cells are screened to select those secreting the antibody against the antigen studied. The selected hybrid cells are cloned and grown in tissue culture or in mouse ascites. Each cloned cell line will produce just one antibody

and should be able to maintain this ability. Antibodies
may be cleaved into fragments that will retain their
binding sites. Reproducibility of the production
process and the purity, potency and safety of the final
product to be radiolabelled is today a major concern.
For radioimaging and for radiotherapy it is necessary
to be able to label with a radionuclide of the desired
properties in a way that the resultant product is
stable and retain the ability to bind to the antigen.
At present the most successful labelling seems to be
the bifunctional chelate labelling with In-111. A
chelate like DTPA is attached to the protein. This
conjugate may be stored and just before use a radionuclide is attached. Iodine labelling is also used but
deiodination <u>in vivo</u> is a main problem. However for
therapeutic purposes this is at present the only radionuclide that has been used. Tc-99m labelling of antibody fragments is also used. Table 7 lists factors to
be considered when evaluating a given product for human
use in order to make sure the product is safe and of a
uniform quality. Contamination with other proteins and
viruses is a major concern. Safety testing in animals
of substances that are foreign proteins to the animal
is also difficult (29). National and international
bodies have prepared guidelines for the testing
requirements (31,32).

<u>Table 7</u> Points to consider before human use of
            monoclonal antibodies. (30)

Immunoreactivity before and after labelling
Homogeneity and purity
Radiochemical purity
Stability
Sterility
Pyrogenes
Safety test in animals
Virus and polynucleotide contamination
Patient skin test

A number of monoclonal antibodies have now been tested
in humans. Some are listed in Table 8. Even a few have
been used in therapy. Results are encouraging. For the
tumor detection it might be necessary to operate with a
mixture of antibodies as the antigenic nature changes
with time.

Table 8. Examples   of   Monoclonal   antibodies   in
preliminary clinical use.

Tc-99m $F(ab')_2$ antimelanomaantibody
In-111   antimyosin
I-123   granulozytspecific MAB 47
I-131   19.9 $F(ab')_2$, anti-CEA $F(ab')_2$ antibody
          (Colon)
I-131   OC 125 $F(ab')_2$ antibody (ovaria)

The human body does also form antibodies against the
mouse proteins and a major step forward would be the
production of human monoclonal antibodies. It is an
area with great prospect but a large amount of research
and development is needed before this new group of
radiopharmaceuticals will   come   in   routine   use.
Particularly the prospects of cancer therapy with
radionuclides or chemotherapeutics attached to
monoclonal antibodies should stimulate the research.

3 RADIOPHARMACEUTICALS FOR POSITRON EMISSION TOMOGRAPHY (PET)

PET is at present a scientific tool. It is not clear
whether it will become a clinical tool with
consequences for treatment of patients (33).   It is
however of great importance as it is the only way that
a large number of biological interesting substances
can be labelled with a radionuclide and made available
for human studies.The most relevant radionuclides are
listed on Table 9 . Many methods have been developed to
label organic molecules like sugars, amino acids and
neurotransmitter receptor ligands. The short half-life
of these radionuclides requires special methods for
synthesis   and   also limits the processes that can
be studied. Research,particularly in brain blood flow
and on receptors, has paved the way for similar
developments with single photon gammaemitting radionuclides as I-123 and Tc-99m and SPECT instrumentation.

Table 9. Characteristics of Radionuclides used for
         PET.

Radionuclide	Half-life (min.)
Carbon-11	20.4
Nitrogen-13	9.96
Oxygen-15	2.07
Fluorine-18	109.7
Gallium-68	68.1
Bromine-75	101

4 RADIOPHARMACEUTICALS FOR THERAPY

Sodium iodide (I-131) is by far the most used radiopharmaceutical for therapy. It is used in the treatment
of hyperthyreodism. This is based on the selective
uptake of iodide by the thyroid (40-50%). It may also
in relatively few cases be used in thyroid cancer.
Y-90 or P-32 colloid treatment of synovitis is also
along with P-32 treatment of polycytaemia   a
classical method but not much used today (34).
Treatment of phaeocromocytoma and neuroblastoma with
I-131 metaiodobenzylguanidine is a new promising area
(23).
The breakthrough in radionuclide therapy may come with
the use of monoclonal antibodies and the search for
useful radionuclides with only β or α radiation
that can be firmly bound to antibodies is underway.

5 RADIOPHARMACEUTICALS: PRODUCTION AND SUPPLY

5.1 Radionuclide production

A large amount of all radionuclide production is based
on reactor irradiation or fission product separation.
Table 1 lists some of the most important radionuclides
and their production method. Industrial production with
cyclotron is of great importance for radionuclides like
Ga-67, In-111, I-123 and Tl-201. The production with
small hospital based cyclotrons is mostly confined to
shortlived positron emitters.

5.2 Radionuclide generators

Tc-99m is mostly prepared at the individual hospital
from a sterile generator system giving a ready for use
solution of sodium pertechnetate. It is based on the
Mo99/Tc99m mother/daughter system. The Mo99 is absorbed
on an aluminium oxide column and the Tc99m eluted with
saline.In some countries a centralised supply of Tc-99m
based on liquid extraction system is working for a few
hospitals or a region (35).
The only other generator system of real importance is
the Rb-81/Kr-81m generator. Rb-81 has a 4.6 hours half
life and is produced by cyclotrons.

5.3   Radiopharmaceuticals: Labelling methods.

Radionuclides may be introduced in a molecule by many different methods. For molecules containing a nuclide that can be replaced by a corresponding radioactive nuclide the methods may be chemical synthesis, biosynthesis or isotopic exchange.For foreign labelling 3 nuclides are most relevant: iodine, indium and technetium.

Iodine. The introduction of iodine in proteins has been known for a long time and iodide must be transferred to its oxidized form. Iodine monochloride labelling, chloramine-T, electrolytic and enzymatic labelling are some of the methods developed (36). The most recent and most lenient method is labelling with Iodo-gen (tetra-chlorodiphenylglycouril). It is used as a "solid phase" dissolved in a small amount of acetone (37).

For exchange labelling of iodine containing radio-pharmaceuticals a fast method based on nucleophilic exchange in the presence of Cu(I) has been developed (38).

Indium. Indium-111 has found several recent applications such as a label for monoclonal antibodies and for cells (39).Antibody labelling follows the general scheme:

Derivation: Attachment of a suitable chelating group to the protein.
Purifaction: Removal of unbound chelate.
Radiolabelling: Attachment of radioindium to protein bound chelate groups.

Well established is the use of cyclic DTPA anhydride for derivatisation.

Indium oxine (In-111 hydroxyquinoline complex) is used for labelling of leucocyttes without loss of their cell-function. In-oxine is lipid soluble and can pass the cell membrane (40).

Technetium. Technetium radiopharmaceuticals are either complexes or foreign labelled compounds. Most labelling methods involve reduction of the pertechnetate to lower oxidation states. The stannous ion is the reducing agent of choice (41). A combination with other reducing agents as gentisic acid may be useful (42) . Many Tc-labelled compounds with "No carrier-added" Tc-99m are probably a mixture of different oxidation states of technetium.

5.4 Supply systems.

The short physical half-life makes the logistics of the supply of radiopharmaceuticals very important. There has in many countries been a development from the use of home made kits for Tc-99m labelled radiopharmaceuticals at individual hospitals towards the use of industrialised products and distribution of as ready-for-use products as possible. Tc-99m generators and preparation kits have become very reliable and simple to use. More complicated procedures are centralised to a few hospitals. It is obvious now that if new labelling procedures such as Tc-99m and Indium-111 labelling of antibodies shall become generally useful, they must be presented in kit form.
Centralised radiopharmacy has been developed in some countries. Such institutions are able to supply hospitals in an area with ready-for-use individual patient doses even of the shorter lived radiopharmaceuticals such as Tc-99m compounds (43). In Denmark a National Isotope-Pharmacy has been combined with a National Control Institute (44).

6. HANDLING OF RADIOPHARMACEUTICALS AT HOSPITALS (GOOD RADIOPHARMACY PRACTICE).

6.1 General.

The quality of a radiopharmaceutical is not just determined by the production methods or the quality control carried out. It may be affected also by the handling and storage, and a quality assurance programme for the whole system must be established (45). The quality of a radiopharmaceutical is defined by setting a standard for the product and the manufacturing process. Good Radiopharmacy Practice is defined as a previously defined manufacturing process carried out and recorded by a trained and qualified staff provided with the necessary facilities including adequate premises, suitable equipment, correct material and approved procedures (35).
The importance of this may be illustrated by the fact that radiopharmaceuticals nearly always must be dispensed individually before administration to the patient due to the continously changing radioactive concentration with time. Hospitals also receive many radiopharmaceuticals as semi-manufactured products in

the form of kits and generators. The final preparation takes place at the hospital and there will be a divided responsibility for the quality of the product between the industrial manufacturer of the kits and generators and the hospital.

To cover the types of handling that take place in hospitals it may be useful to describe four types of radiopharmaceuticals:

Ready-for-use radiopharmaceuticals:
These radiopharmaceuticals contain radionuclides with a sufficiently long half-life to allow distribution from the manufacturer to the hospital of a product in a ready-for-use form. The volume of the radiopharmaceutical to be administered can only be determined after calculation of the radioactive decay.

Radiopharmaceuticals prepared from semi-manufactured products:
The use of a radionuclide such as Technetium-99m with a half-life of 6 hours has led to the development of radionuclide generators and preparation kits that can be used for a longer period of time. It has proved to be practical to extend drug control to these products as part of approval of the safety and efficacy of radiopharmaceuticals given to patients.

Radiopharmaceuticals prepared directly in relation to the administration:
The radionuclide may have such a short half-life that administration must take place immediately after preparation. Examples are radioactive gases or solutions of these prepared by cyclotrons or with the help of radionuclide generators containing a longer lived parent-radionuclide.

Radiopharmaceuticals based on samples from patient:
(Autologous labelled products): Cells or plasma proteins from a patient may be labelled with radioactivity before re-administration to the same patient.

6.2 Elements of Good Radiopharmacy Practice.

Personnel. The availability of a sufficiently well-trained and qualified staff is one of the most important elements of Good Radiopharmacy Practice. The duties and responsibilities of each individual should be clearly defined and described in writing.

Premises. The design of rooms for the preparation of radiopharmaceuticals has two main purposes: to protect the operator from radioactivity, and to protect the product from contamination by the operator and the rest of the environment. This is achieved by a design where cleaning is facilitated by the construction and type of

furnishing used, where the air is continously exchanged
with clean air and where a system of barriers is built
up around the product to prevent very large differences
in contamination levels. Such barriers may, for
example, consist of a closed vial, a contained work
station with laminar air-flow, and a ventilated
laboratory room within a building. A source of
contamination is brought in whenever the operator or
his hands enter the system. The risk is minimized by
dressing the operator in suitable protective clothing,
which also serve to protect him from the product. The
ventilation system also serves to prevent the spread of
radioactivity. In general, hygienic and radiation
protection requirements for the design of laboratories
are of the same nature. As far as ventilation systems
and the installation of sinks are concerned, there may
be opposite considerations and special design has to be
used.

Documentation. Documentation is absolutely
indispensable in a quality assurance system. The main
purposes are: to define and describe the system of
control to reduce the risk of error that purely verbal
communication introduces and to make it possible to
trace an individual product back through the whole
process including the utilization and disposal of
starting materials, packaging materials and finished
products.

All documentation material must be kept up-to-date and
instructions, master formulas, etc. must be written in
such a way that no doubt can arise about which issue is
valid and which changes have been made. All materials
should be dated and should be authorized by the
relevant competent persons.

Quality control. A quality control system must be
established for each radiopharmaceutical prepared. It
may consist of complete analytical testing of each
individual batch or may rely on control of the method
by continuous surveillance of certain parameters
independent of the release for use of individual
batches. Such a system will always be supplemented by a
few checks on the individual batch.

6.3 Handling of radiopharmaceuticals at hospitals.

General aspects. Most handling of radiopharmaceuticals
at hospitals is very simple, such as dispensing patient
doses from stock solutions (handling of ready-for-use
radiopharmaceuticals) or simple preparations of short-
lived radiopharmaceuticals from radionuclide generators
and preparation kits, in particular preparation of
sodium pertechnetate (Tc-99m) injections and

preparation of Tc-99m-labelled radiopharmaceuticals. In principle GRP as described is used, but there are reasons for a simpler system as a greater part of the work has been done before the radiopharmaceuticals or semi-finished products arrive at the hospital.

Ready-for-use radiopharmaceuticals. The laboratory facilities required may be rather limited when dispensing is carried out at an individual hospital for immediate use. Before a radiopharmaceutical is ordered, a decision must be made as to what specifications have to be fulfilled.

No quality control will normally have to be performed with ready-for-use products. This has been observed by the manufacturer. Visual inspection, control of paperwork and control of the activity of individual patient doses is all that will be needed.

Preparation of sodium pertechnetate (Tc-99m) injections from Tc-99m generators. The starting material is a Tc-99m generator produced with the intention that the eluted product will fulfil the requirements of a pharmacopoeia or a similar specification. The generator is delivered with all necessary accessories for elution including the sodium chloride solution. No other solutions should be used. The radiation protection aspects and hygienic conditions must be considered when locating a Tc-99m generator.

Quality control should include properties that may be affected by transportation of the generator and operation at the hospital: eluting yield, Mo-99 in eluate and control of aseptic technique.

Tc-99m-labelled radiopharmaceutical prepared from preparation kits. Preparation is carried out using sodium pertechnetate (Tc-99m) injection and preparation kits. It is assumed that such kits are produced in accordance with generally good pharmaceutical manufacturing practice and according to fixed specifications either in pharmacopoeias or in registration documentation.

As far as Tc-99m prepared from generators is concerned, responsibility is shared between the manufacturer of the preparation kit and the person responsible for preparing the Tc-99m-labelled radiopharmaceutical at the hospital.

Generators and kits may originate from different manufacturers, as not all kit manufacturers produce generators and not all generator manufacturers produce all types of kits. Thus, there may be the problem of compatibility between generators and kits. Information should be requested from the manufacturer but the final responsibility must, unfortunately, rest with the hospital, unless kit and generators from the same manufacturer are used.

All kit procedures are performed in closed systems and the resulting products are used within one working day. Most kits require only a single-step procedure. The requirements for hygienic conditions and radiation protection are therefore similar to those required for dispensing ready-for-use radiopharmaceuticals. Manufacturers give full instructions on preparation details. They should be strictly followed and interpreted to the technical staff performing the daily procedures.

Analytical quality control of routine preparation when using Tc-99m kits and generators should not be needed. It is, however, useful to be able to perform a simple radiochemical purity test for use in trouble-shooting whenever strange imaging results are obtained.

In order to be able to use such a test, it has to be practiced. It is therefore recommended that it should be used whenever a new type of Tc-99m generator is introduced or a new technician is acquainted with his new work, and possibly at regular intervals.

Labelling of red blood cells and other biological samples with the help of preparation kits. Labelling of samples from patients with radionuclides presents the same problems as the preparation of other radiopharmaceuticals, and the same principles as already described should be used.

There are two complicating factors:

a) Procedures are often multi-steps and may include centrifugations;

b) Patient samples may contain infectious material, such as hepatitis virus, presenting an added risk to the operator and the risk of cross-contamination between samples.

Although preparation takes place in closed systems, the added infection risk requires that the procedure be separated from other radiopharmaceutical procedures. The products are growth media for microorganisms and the requirement of a good aseptic technique is therefore stressed.

As far as possible, work should take place in a downward-flow laminar air-flow (LAF) work station situated in a separate room. If work has to be carried out in the same room as normal radiopharmaceutical preparation work in a closed system, it is considered necessary to have a separate LAF unit or a suitable box structure for handling these biological materials.

The simplified quality control testing may have to include, apart from the problem of labelling yield, consideration of possible damage to the biological sample. Control of the aseptic technique must be performed.

7   SAFETY AND EFFICACY

7.1   General

A new pharmaceutical is not introduced into general use before it has been tested in a way that an extensive knowledge about its efficacy and safety is available. After the design of a new chemical compound, methods are developed for its description, identification and determination. A stable pharmaceutical form and a production system are developed and a clinical trial through phase 1 to 3 is carried out. At a later stage, phase 4, a continued surveillance is carried out by the manufacturer and maybe also by the registration authorities.

In general most of the requirements concerning documentation on the safety and efficacy of a non-radioactive drug will also be relevant to radiopharma-ceuticals. It must, however, be recognized that in many cases a radiopharmaceutical will have no measurable pharmacodynamic effect as the purpose is only to utilize its pharmacokinetic properties for diagnostic purposes. Radiopharmaceuticals for therapy have however, as already described, the biological effect of radiation as a general property. In addition to the general requirements on documenting safety and efficacy, including radiopharmaceutial parameters it is necessary to describe radiation hygiene parameters, related to the patient and to the staff handling such material. An extra dimension is hereby added in the weighing of risk and benefit. The radiation risk may, in contrast to other risks be expressed in the same unit for different radiopharmaceuticals, whereby a direct comparison between different radiopharma-

ceuticals may be possible. It is necessary for
documentation on a radiopharmaceutical to cover the
entire production and handling process up to
administration to the patient, as handling may
seriously effect its safety and efficacy.
Radiopharmaceuticals have changing composition with
time due to radioactive decay. The physical half-life
of the radionuclide is often so short that the final
preparation must take place immediately before
administration to the patient. This can be made
possible by the use of semi-manufactured products such
as radionuclide generators and preparation kits. To be
able to establish the safety and efficacy of radiopharmaceuticals it is also necessary to include
standardization of generators and kits and other semi-
manufactured products. In other cases, such as the use
of ultra short-lived radioactive gases, the only means
of standardization is a reproducible production
process, as the radiopharmaceuticals go direct from the
generator or the cyclotron to the patient. Another area
where the establishment of specifications may require
special attention is the labelling of samples from the
patient with a radioactive substance before re-
administration to the patient (46,47)

7.2   Standards for radiopharmaceuticals

A specification (standard) for a radiopharmaceutical
either in a pharmacopoea monograph or in a registration
file should at least contain:
- radionuclide purity
- specific activity
- radiochemical purity
- radioactive concentration
- pH
- isotonicity
- particle size (if relevant)
- sterility
- apyrogenicity
- stability
- packaging material.
For preparation kits specifications must be given both
for the kit and the final labelled product.

7.3 Safety testing and documentation

The extent of toxicity testing will depend on, and must be designed to relate as closely as possible to, the intended clinical use of the radiopharmaceutical. For a radiopharmaceutical intended for single dose administration it is normally required that it has been tested for single and repeated dose (2 weeks) toxicity in at least two animal species. New chemical entities must be evaluated for mutagenic and carcinogenic potential.

Pharmacodynamic and pharmacokinetic properties must be described in animals and humans particularly to provide the background for radiation dosimetric calculations. For monoclonal antibodies there may be special problems of virus and foreign protein contamination.

7.4 Efficacy studies

For non-radioactive pharmaceuticals very detailed studies are required in relation to the claimed indications for the drug.

A radiopharmaceutical is often part of a diagnostic system, and it is not useful to try to fix very limited indications. However, it is relevant that clinical trials establish the safety in humans and the diagnostic/therapeutic efficiency including sensitivity and specificity.

8 REGULATORY ASPECTS

8.1 Regulations

Radiopharmaceuticals are regulated both from a drug regulatory and from a radiation protection point of view. Radiation protection regulations are in force in most countries, while drug regulations on radiopharma-ceuticals are not yet introduced in all countries. The purpose is of course to make sure that only safe and effective radiopharmaceuticals of a uniform quality are in use. Documentation is therefore required for the safety and efficacy as described in section 7.

A regulatory system would normally comprise the following:

- setting of standards
- licensing of production methods and/or facilities
- licensing of hospital preparations
- licensing of distributors
- licensing of centralized radiopharmacy
- official control and inspection

- licensing or reporting of clinical trials.
National regulations on these items exist in a number
of European countries (48,49). Directives on radiopharmaceuticals are now in preparation within the EEC
(50). Special draft guidelines for monoclonal
antibodies have been prepared (31). A set of Nordic
guidelines for registration of radiopharmaceuticals are
under preparation (51). Standards in Europe are set in
European Pharmacopoea. A list of monographs is given in
table 10.

Table 10   Specifications for radiopharmaceuticals in
            the European Pharmacopoea

Aqua tritiatae (H-3) solutio iniectabilis
Auri colloidalis (Au-198) solutio iniectabilis
Chlormerodrini (Hg-197) solutio iniectabilis
Chromii edetatis (Cr-51) solutio iniectabilis
Cyanocobalamini (Co-57) solutio
Cyanocobalamini (Co-58) solutio
Hydrargyri dichloridi (Hg-197) solutio iniectabilis
Krypton (Kr-85) solutio iniectabilis
Natrii chromatis (Cr-51) solutio sterilis
Natrii iodidi (I-125) solutio
Natrii iodidi (I-131) solutio
Natrii iodohippurati (I-131) solutio iniectabilis
Natrii pertechnetatis (Tc-99m) fissione formati solutio
iniectabilis
Natrii pertechnetatis (Tc-99m) sine fissione formati
solutio iniectabilis
Natrii phosphatis (P-32) solutio iniectabilis
Rhenii sulfidi colloidalis et technetii (Tc-99m)
solutio iniectabilis
L-Selenomethionini (Se-75) solutio iniectabilis
Stanni pyrophosphatis et technetii (Tc-99m) solutio
iniectabilis
Stibii sulfidi colloidalis et technetii (Tc-99m)
solutio iniectabilis
Sulfuris colloidalis et technetii (Tc-99m) solutio
iniectabilis
Technetii macrosalbi (Tc-99m) suspensio iniectabilis
Xenoni (Xe-133) solutio iniectabilis.

When licensing industrial and hospital facilities, it
is important to correlate this task with the work of
the regulatory authorities for radiation protection in
order not to have two agencies putting up conflicting
requirements.

The official control may be organized in many different ways, but it is very important that an overall surveillance programme be established. This may consist of decentralized or centralized analytical control, inspection and reporting systems for adverse reactions and drug defects. A central evaluation of the results is indispensable.

8.2 Quality assurance

The quality of the radiopharmaceutical, after administration to the patient, can be observed by physicians during and after the nuclear medical procedure. During the interpretation of images the physician should be aware of false biological distributions.
Radiopharmaceutical defects detected by quality control procedures or by similar observations should be reported to the manufacturer as well as to any relevant national, regional, or international committee which may have been set up to monitor and record such defects.

Table 11   Drug defects reported in 1986 from hospitals in Denmark (7), The Netherlands (8) and United Kingdom (62).

	Apperance	Total Radio-activity	Radio-chemical purity	Elution effi-ciency	Particu-late contami-nation	Bio-distri-bution	Other	Total
Tc-99m Generator				37			1	38
-DMSA						3		3
-DTPA			1					1
-HMDP						2		2
-HSA colloid							2	2
-MAA			1		1	7		9
-MDP	1		1		1	3		6
-Microspheres	1							1
-Phyrosphosphate						1		1
-RBC label						2	1	3
-Tin colloid						2		2
In-111 Oxin							1	1
I-123 -MIBG						2		2
I-123 -NaI							1	1
I-131 -Fibrinogen					3			3
I-131 -NaI		2						2
Total	2	2	3	37	5	22	6	77

Unexpected adverse reactions, such as pyrogenic, vasovagal and allergic ones, should also be reported to the manufacturer and to any relevant national, regional, or international committee.
Such reporting systems are considered to be of primary importance in order to obtain an early detection of any defect or adverse reaction.

In Europe the European Joint Committee on Radiopharmaceuticals of the European Association of Nuclear Medicine has established a reporting system with a secretariat in Denmark (52).
Table 11 gives a list of drug defects reported from Denmark, Holland and United Kingdom in 1986. From the Danish Isotope-Pharmacy receiving control programme it was shown that in 2.2% of all shipments a defect was found that required communication with the supplier.

The frequency of adverse reactions from radiopharmaceuticals is low compared to most other groups of drugs. The reactions are most often mild and of short duration illustrated by the fact that only 6 patients out of a reported 24 required treatment. Table 12 gives the reports received in 1986 in the European system. It should be stressed that only reports from Denmark, Holland and the United Kingdom were received.

Table 12   Summary of reports on adverse reactions 1986

		Number of reports	Types of reaction	Number of patients requiring treatment
Tc-99m	Antimony Sulphur collorid	1	a	
	DMSA	1	a	
	DTPA	7	a,b	2
	HMDP	1	d	
	MDP	4	a,b,c	2
	RBC	1	a	
	Tin collorid	1	a	
Cr-51	EDTA	1	a	
Ga-67	Citrate	1	d	
Se-75	Selencholesterol	1	c	1
I-123	Hippuran	1	b	
I-123	MIBG	2	a,d	
I-123	Sodium Iodide	1	a	
I-131	Sodium Iodide	1	d	1
Total		24		6

Types of reactions: a: Allergic, b: Vasovagal, c: Local effects, D: Other effects.

REFERENCES

1. G.M.Pohost,N.M.Alpert,J.S.Ingwall,H.W.Strauss,Thallium Redistribution: Mechanism and Clinical Utility, Sem.Nucl.Med.1980, 10,70.
2. S.M.Larson,Mechanism of localization of Gallium-67 in Tumors,Sem.Nucl.Med.1978,8,193.
3. M.W.Merrich,Essentials of Nuclear Medicine, Churchill Livingstone,Edinburgh,1984.

4. R.L.P.van den Bosch,J.J.M.de Goeij,J.F.W.Tertoolen, M.W.M.Tielemans, Radionuclidic purity of I-123 productions,J.App.Rad.Isotopes, 1978, 29,343-44.

5. M.D.Short,Radiopharmaceuticals for lung ventilation studies in Radiopharmacy & Radiopharmaceuticals, A.E.Theobald(editor), Taylor & Francis, London 1985

6. W.M.Burdi,P.J.Sullivan,F.E.Lomas,V.A.Evans,C.J.McLaren, R.N.Arnot, Lung ventilation studies with Technetium-99m Pseudogas,J.Nucl.Med.1986,27,842-46.

7. L.Bergqvist,S-E.Strand,B.R.Persson, Particle Sizing and Biokinetics of Interstitial Lymphoscintigraphic agents, Sem.Nucl.Med.1983,13,9-19.

8. S.C.Srivastra,L.R.Cherva,Radionuclide-labelled Red Blood Cells:Current status and Future Prospects.Sem.Nucl.Med.1984,14,68.

9. A.M.Peters,H.J.Danphure,S.Osman,R.J.Hawker,B.L.Henderson,H.J.Hodgson,J.D.Kelly,R.D.Neirinckx,J.P.Lavender, Clinical experience with Tc-99m-hexamethylpropylene amine oxine for labelling leucocytes and imaging inflammation. Lancet 1986,ii:946.

10. K.Kristensen,Biodistribution in rats of Tc-99m-labelled human serum albumin,Nucl.Med.Comm.1986,7,617.

11. L.R.Chervu,M.D.Blaufox,Renal Radiopharmaceuticals-An Update. Sem.Nucl.Med.1982,12,224.

12. W.C.Klingensmith,A.R.Fritzberg,W.M.Spitzer,D.L.Johnson ,C.C.Kuni, M.R.Williamson,G.Washer,R.Weil,Clinical evaluation of Tc-99m NN′-Bis (mercaptoacetyl)-2,3-Diaminipropanoate as a replacement for I-131 Hippurate:Concise communication,J.Nucl.Med.1984,25, 42-48.

13. A.Taylor,D.Eshima,A.R.Fritzberg,P.E.Christian,S.Kasina, Comparison of Iodine-131-OIH and Technetium-99m MAG₃ Renal Imaging in Volunteers.J.Nucl.Med.1986, 27,795-803.

14. A.R.Fritzberg,S.Kasina,D.Eskima,D.L.Johnson, Synthesis and Biological Evaluation of Technetium-99m MAG₃ as Hippurane Replacement.J.Nucl.Med.1986,27,111.

15. G.Subramanian,J.G.McAfee,R.F.Schneider, Structure distribution relationship in the design of Tc-99m radiopharmaceuticals, Safety & Efficacy of radiopharmaceuticals,K.Kristensen,E.Nørbygaard (editors), Nijhoff, The Hague 1984.

16. L.R.Chevu,A.D.Nunn,M.D.Loberg,Radiopharmaceuticals for Hepatobilaory Imaging,Sem.Nucl.Med.1982,12,5.

17. P.J.Ell, Cerebral blood flow,Nucl.Med.Comm.1987,8,453.

18. P.F.Sharp,F.W.Smith,H.G.Gemmel,D.Lyall,N.T.S.Evans, D.Gvozdanovse,J.Davidsen,D.A.Tyrrell,R.D.Pickett,R. D.Neirincx, Technetium-99m HM-PAO stereoisomers as potential agents for imaging regional blood flow: Human volunteer studies.J.Nucl.Med.1986,27,171.

19. R.D.Neirincx,R.C.Harrison,A.M.Forster,J.F.Barke, A.R. Andersen,N.A.Lassen,A model for the in vivo behaviour of Tc-99m d.1 HM-PAO in man (abstr.)J.Nucl.Med.1987, 28,559.

20. B.Johannsen,Newer Technetium-99m radiopharmaceuticals in Safety & Efficacy of Radiopharmaceuticals 1987, K.Kristensen,E.Nørbygaard,(editors),Nijhoff,Dordrecht, 1987,21.

21. B.D.Bok,V.Scheffel,H.W.Goldfarb,H.D.Burns,S.E.Lever, D.F.Wong,A.Bice,H.N.Wagner,Comparison of Tc-99m complexes (NEP-DADT,ME-NEP-DADT and HMPAO) with I-123 IAMP for brain SPECT imaging in dogs. Nucl.Med.Comm.1987,8,631.

22. K.McKusik,B.L.Holman,A.G.Jones,A.Davison,P.Rigo, V.Sporn,H.Vosberg,J.Moretti,Comparison of 3 Tc-99m isonitriles for detection ischemic heart disease in humans (abstr),J.Nucl.Med.1986,27,878.

23. W.H.Beierwalters,Clinical Application of I-131 labelled Metaiodobenzylguanidine, 1987 yearbook of Nuclear Medicine, P.B.Hoffer,editor,Yearbook Medical Publishers,Chicago,1987,17.

24. S.D.Sarkar,W.H.Beierwalters,R.D.Ice,G.P.Basmadijan, K.R.Hetzel,W.P.Kennedy,M.M.Mason,A new and superior adrenal scanning agent (NP59),J.Nucl.Med.1975,16,1038.

25. A.L.Hawkins,K.E.Britton,B.Sharpiro,Selenium-75 Seleno-methylcholesterol:A new agent for quantitative functio-nal scintigraphy of adrenals:Physical aspects.Brit.J.-Radiol.1980,53,883.

26. H.R.Schelbert,Current status and prospects of new radionuclides and radiopharmaceuticals for cardio vascular nuclear medicine.Sem.Nucl.Med.1987,17,145.

27. M.R.Kilbourn,M.R.Zalutsky,Research and clinical poten-tials of receptor based radiopharmaceuticals.J.Nucl.-Med.1985,26,655.

28. J.Zeuthen,Monoclonal antibodies and their radionuclide conjugates:Practical and regulatory aspects in Safety & Efficacy of radiopharmaceuticals 1987, K.Kristensen,-E.Nørbygaard,(editors),Nijhoff,Dordrecht, 1987,129.

29. O.Svendsen,H.B.Christensen,P.Juul,J.Rygaard,Models for Safety testing of immuno reactive pharmaceuticals in Safety & Efficacy of Radiopharmaceuticals 1987,K.Kristensen,E.Nørbygaard,(editors),Nijhoff, Dordrecht,1987,105.

30. G.Burragi,The joint task group on clinical utility of labelled antibodies of the SNME & the ENMS,The Int.J.Biol.Markers,1986,1,147.

31. On the production and quality control of monoclonal antibodies of murine origin intended for use in man.Notes to applicants for Marketing authorisation.Commission of the European Committies,Bruxelles,1987,III/859/86.

32. Points to consider in the manufacture and testing of monoclonal antibody products for human use (1987),U.S.Department of Health & Human Services. Food & Drug Administration, Washington DC, 1987.

33. M.Reivich,A.Alavi(eds),Positron Emission Tomography,A.R.Liss,N.Y.1985.

34. E.L.Saenger,J.G.Kereiakes,V.J.Sodd,R.David,Radiotherapeutic Agents:Properties,Dosimetry and Radiobiological Considerations,Sem.Nucl.Med.,1979,9,72.

35. K.Kristensen,Preparation and control of radiopharmaceuticals in hospitals.,IAEA,Vienna,1979.

36. M.Argentini,Labelling with iodine. A review of literature. Federal Inst. for Reactor Research, Würenlinge, Switzerland,1982,319p.

37. A.P.Richardson,P.J.Mountford,A.C.Baird,T.C.Richardson, A.J.Coakley,An improved Iodogen method of labelling antibodies with I-123,Nucl.Med.Com.1986,7,355.

38. J.Mertens,W.Vanrycheghem,M.Gysemans,J.Essels,E.Findu-Panek,L.Carlsen,New fast preparation of I-123 labelled radiopharmaceuticals,Eur.J.Nucl.Med.1987,13,380.

39. S.Mather,Labelling with Indium-111 in Safety & Efficacy of Radiopharmaceuticals,1987,K.Kristensen,E.Nørbygaard, (editors),Nijhoff,Dordrecht,1987,51.

40. H.J.Danpure,S.Osman,Specifications and quality control methods for labelled cells. in Safety & Efficacy of Radiopharmaceuticals 1987,K.Kristensen,E.Nørbygaard,(Editors),Nijhoff, Dordrecht,1987,161.

41. E.Deutch,Recent developments in Technetium Chemistry as applied to the generation of new Tc-99m radiopharmaceuticals in Radiopharmacy & Radiopharmacology yearbook 1,P.H.Cox,editor, 1985,Gordon & Breach, New York, 1985,69.

42. E.Sundrehagen:Labelling methods with Technetium-99m in Safety & Efficacy of Radiopharmaceuticals 1987-K.Kristensen,E.Nørbygaard(editors),Nijhoff,Dordrecht,-1987,67.

43. E.Vaughn,Pharmacy enlarges role in radioactive drugs,Drug Topics,1983,76.

44. K.KristensenQuality Control of Radiopharmaceuticals in Medical Radionuclide Imaging 1980(Proc.Symp.Heidelberg 1980),vol.2,IAEA,Vienna,1981,59.

45. Quality Assurance in Nuclear Medicine,World Health Org.,Geneva,1982.

46. K.Kristensen,E.Nørbygaard,(editors),Safety & Efficacy of radiopharmaceuticals,Nijhoff,Dordrecht,1984.

47. K.Kristensen,E.Nørbygaard,(editors),Safety & Efficacy of radiopharmaceuticals 1987,Nijhoff,Dordrecht,1987.

48. K.Kristensen,Regulatory aspects of radiopharmaceuticals in Radiopharmaceuticals and labelled compounds 1984,IAEA,Vienna,1985,383.

49. K.Kristensen,Licensing of radiopharmaceuticals in European countries 1982 in Safety & Efficacy of radiopharmaceuticals,Nijhoff,Dordrecht,1984,299.
50. European Economic Community,Proposed directive for radiopharmaceuticals. In preparation 1987.
51. Nordic Council on Medicines,Radiopharmaceuticals,Drug applications,Nordic guidelines,In press, Uppsala,1988.
52. Joint Committee on Radiopharmaceuticals of the European Nuclear Medicine Society and the Society of Nuclear Medicine-Europe,Eur.J.Nucl.Med.1984,9,308.

9
Isotopes in Molecular Biology

P.S.G. Goldfarb

MOLECULAR TOXICOLOGY GROUP, DEPARTMENT OF BIOCHEMISTRY, UNIVERSITY OF SURREY,
GUILDFORD, SURREY, GU2 5XH, UK

1 INTRODUCTION

Molecular biology was first defined as the study of the structure of biological macromolecules. In its modern context this has now been expanded to include the analysis of macromolecular function, synthesis and regulation. The ability to study these aspects of cellular growth and development has resulted from improvements in techniques to isolate and purify macromolecules, such as nucleic acids or proteins, coupled with an increasing sophistication in detection systems. It is now possible to carry out in vitro investigations on the purified molecules rather than on whole cells or sub-cellular fractions. In particular, the purification of proteins to homogeneity has enabled the molecular biologist to use enzymes as in vitro biological catalysts. Thus specific modification or dissection of larger biological polymers can also be performed to facilitate the analysis of their structure and function.

2 THE CENTRAL DOGMA

The central theme of biology is that information flow in the living cell proceeds as shown in Figure 1.

DNA ⟶ RNA ⟶ Protein

Figure 1 Basic dogma of biology

What this says is that the programme for cellular growth and development resides in the macromolecule deoxyribonucleic acid (DNA). This programme is functionally divided up into informational packages known as the genes, which supply the information for the production of cellular components. The instructions, for the production of the structural and catalytic components of the cell, the proteins, are transferred by an informational ribonucleic acid molecule, messenger RNA, to the protein synthesising apparatus of the cell.   Here proteins are made in conformation to the instructions laid down in the DNA.

The three types of biological macromolecule of most interest to the Molecular Biologist are thus DNA, RNA and protein. I will deal with each of these in turn and demonstrate how the use of isotopes has been essential to the elucidation of their structure and function.

3 THE STRUCTURE AND FUNCTION OF DNA

Structure and Replication

DNA is a polymer made up of the four nucleosides deoxyadenosine, deoxyguanosine, thymidine and deoxycytosine. The nucleotides are joined in a linear fashion <u>via</u> 5'-3' sugar-phosphate bonds to form the backbone of the molecule. In fact, DNA is a double-stranded molecule in which the two nucleotide chains are wound helically around each other and held together by hydrogen bonds. These hydrogen bonds are formed between the base rings of the nucleotides, adenine pairing with thymidine, and guanine with cytosine. As described by Watson and Crick in 1953 (1) this structure provides a means for replication and copying of the molecule.

The cellular strategy for DNA replication was demonstrated by Meselson and Stahl in 1957 (2) using the ^{15}N isotope of nitrogen to label DNA strands. This enabled template strands which were labelled with ^{15}N to be separated from newly-synthesised strands containing ^{14}N by isopyric centrifugation in cesium chloride gradients.   The experiments clearly demonstrated, as shown in Figure 2 that DNA replication was semiconservative and that the daughter molecules were hybrids of old and new strands.

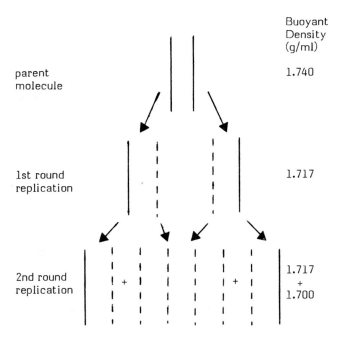

Figure 2 Semiconservative replication of DNA

—— = DNA containing ^{15}N, - - - - - = DNA containing ^{14}N

Genetic Engineering

Such experiments, however, and subsequent work on the mechanism of replication treated the cellular DNA as a homogeneous material. In fact DNA is heterogeneous, as would be expected for an informational molecule whose information content is defined by the sequence of nucleotide bases. The base composition of DNA therefore varies along its length and further investigation of the structure and function of individual regions or genes was restricted until the advent of genetic engineering in the early 1970s. The breakthrough came with the isolation of the restriction enzymes which can cut DNA in a precise fashion (3) and DNA ligase, an enzyme which can join double stranded DNA fragments together. Thus it was now possible to isolate individual genes or precise fragments of DNA and to clone them in bacteria (4,5). The bacteria can then be propagated on a large scale.

There are four main advantages to this technology. First, genes or gene fragments can be obtained uncontaminated by other DNA species. Second, the cloned DNA can be prepared in large quantities for use as an analytical reagent to detect DNA containing complementary nucleotide sequences. Third, each cloned DNA fragment is specific in nucleotide sequence, and hence biological function, not only to the species of organism from which it was derived, but also to the individual for which the original DNA sample was obtained. Fourth, if the appropriate regulatory sequences are attached to the cloned gene, then the recipient bacteria can be tricked into producing large quantities of the protein specified by the informational nucleotide sequence in the 'foreign' gene. The use of individual genes or DNA fragments as analytical probes required the development of methods for 'tagging' the DNA so that it could be used in assays of high-sensitivity and specificity.

Radioisotopic Labelling of DNA

The most useful isotopes for labelling DNA in vitro have proved to be high energy β emitters such as ^{32}P or ^{35}S. These radionuclides can be incorporated into DNA molecules by a variety of techniques using purified cellular enzymes as in vitro catalysts. In particular the use of the techniques of 'nick translation' (6) using $[\alpha-^{32}P]dCTP$ (Figure 3) and 'end labelling' (7) using $[\gamma-^{32}P]ATP$ (Figure 4) have proved the most useful in producing DNA probes of high specific activity. In some instances precursors in which the α or γ phosphate is substituted by ^{35}S are used in preference as this gives less scatter on autoradiographs .

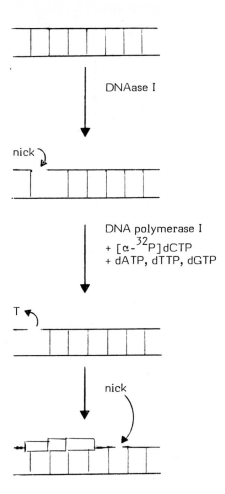

Figure 3 Nick translation. DNA is nicked by DNAase I at low concentration and 14°C. Nucleotides are progressively removed and replaced by DNA polymerase I. If a radioactive nucleotide is included as substrate this will become incorporated, thus tagging the DNA.

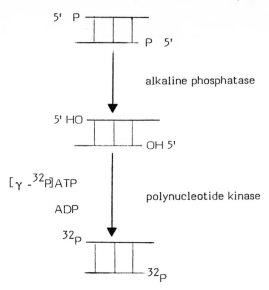

<u>Figure 4</u> End labelling. The 5' terminal phosphate groups of the DNA are removed by alkaline phosphatase. They are subsequently replaced by ^{32}P phosphate from [γ-^{32}P]ATP using the enzyme polynucleotide kinase.

Use of Radio-Labelled DNA Probes

The use of DNA probes to analyse gene structure and function depends on the ability of the probe to hybridise to homologous sequences in target DNA or RNA by hydrogen bonding (Figure 5). The probe and target are first denatured to produce single stranded molecules.

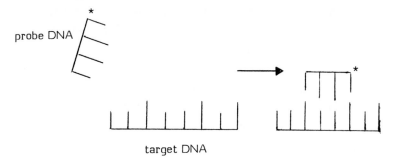

probe DNA

target DNA

Figure 5   Hybridisation of a radioactively labelled probe DNA to target nucleotide sequences. The probe will only bind stably to its exact complementary sequence according to the base-pairing rules (A with T, G with C).

Usually the target DNA is immobilized by attachment to a solid support such as nitrocellulose. After washing of the filter, hybridization of the labelled probe can be detected by autoradiography using X-ray film.

DNA probes have been used for a wide variety of purposes including antenatal diagnosis of genetic disease or chromosome abnormality (8), DNA fingerprinting for forensic purposes (9), and the detection of infectious agents (10). They are a far more precise tool than any other method including the use of immunological procedures, and tests can be performed on far smaller amounts of sample.

4   RNA STRUCTURE AND FUNCTION

RNA Structure and Synthesis

RNA is a single stranded ribonucleotide polymer produced by the copying of DNA by the enzyme RNA polymerase. Three kinds of RNA exist in the cell: ribosomal RNA, transfer RNA, and messenger RNA. All three molecules can take up extensive secondary folding by the formation of intra-chain base pairing, the extent of folding to some extent reflecting the function of the molecules i.e. structural, catalytic or informational respectively. The synthesis of RNA is tightly controlled by the cell, messenger RNA (mRNA) synthesis in particular being regulated by cellular responses either to the availability of substrates for growth or energy production, or in a multicellular organism to the metabolic function of different tissues. This is a reflection of the fact that not all genes in DNA are used at the same time to provide information for the synthesis of proteins.

Use of Radiolabelled Probes to Analyse mRNA

As a mRNA is complementary in nucleotide sequence to the gene from which it was transcribed, the DNA fragments of the gene labelled with ^{32}P by nick translation or end labelling can be used to determine mRNA amount and size (11). More recently new procedures using radiolabelled RNA probes have been devised (12) as these give a greater yield of probe and hence enable a stronger signal to be obtained in the hybridization reaction. Here an in vitro reaction containing purified RNA polymerase is used to copy the DNA gene fragment and incorporate $[\alpha-^{32}P]ATP$ into the growing RNA chain (Figure 6).

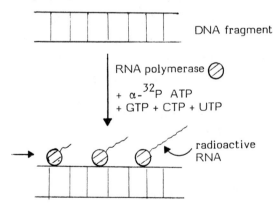

Figure 6   Use of RNA polymerase to produce radiolabelled RNA. The system is designed so that labelled RNA (riboprobe) complementary to the cellular mRNA is produced thus enabling it to hybridize by base pairing.

As with DNA analysis the target mRNA is usually attached to nitrocellulose or nylon for ease of handling and detection. Again autoradiography is used to detect hybrid formation.

The analysis of RNA structure and function using radiolabelled probes has greatly increased our understanding of how gene expression is regulated and the regulatory factors involved in controlling cellular phenotype.

5 PROTEIN SYNTHESIS

Protein Structure and Function

Proteins are the structural and catalytic components of cells. They are long chain polymers made up of amino acids condensed <u>via</u> the -CO-NH- peptide bond. They were in fact the first macromolecules to be studied at the molecular level. Proteins come in a variety of shapes and sizes, some showing extensive folding to produce globular molecules, others such as the fibrous proteins, demonstrating a more extended structure. Many proteins are made up of a number of polypeptide subunits aggregated together to form a functional complex.

Cellular Synthesis of Proteins

The sequence of amino acids in a polypeptide chain is determined by the information (nucleotide sequence) in its corresponding gene, and hence mRNA. Synthesis is performed by a sub-cellular particle, the ribosome, which has the ability, in conjunction with transfer RNA, to read the nucleotide sequence in a mRNA and hence condense the amino acids together in the correct order. This is in accordance with the universal genetic code for the translation of nucleotide sequence into amino acid sequence. Each amino acid is coded for by a precise sequence of three nucleotides, a triplet codon, the mRNA thus being read in group of three nucleotides.

Radiolabelling of Proteins

Protein molecules can be labelled in vivo and in vitro, using radioactively labelled amino acids. For in vivo work ^{3}H or ^{14}C labelled leucine is used as this is relatively stable. For in vitro labelling ^{35}S-methionine is often used as smaller amounts of protein can be detected following in vitro synthesis. ^{125}Iodine has also been used to label proteins directly. The hazards associated with the use of this volatile substance has however resulted in its use being limited to situations where other methods of labelling cannot be used. For some proteins which undergo modification, e.g. phosphorylation, glycosylation, labelling with ^{32}P from ^{32}PO$_4$ or [γ-^{32}P]ATP, or with ^{14}C glucosamine, can be performed in vivo or in vitro. Following radiolabelling the proteins can be separated by electrophoresis in polyacrylamide gels and detected by autoradiography (13). For more precise analysis of the synthesis of an individual protein for which an antibody is available, immunoprecipitation prior to gel electrophoresis can be performed.

The application of these techniques has enabled molecular biologists to obtain a greater understanding of the mechanisms regulating the synthesis of cellular proteins. Protein chemists, cell biologists, and immunologists have also benefitted from the use of radioisotopes to label proteins. New insights have thus been obtained into the structure and function of specific proteins in both the individual cell and the whole organism.

6 NON-ISOTOPIC LABELLING OF MACROMOLECULES

While radioisotopes have been of great use in basic molecular research as an aid to understanding the synthesis, structure and function of macromolecules, the use of high energy β-emitters in the clinical or routine analytical laboratory has presented problems, particularly with regard to health hazards and the short half life of the radionuclides.

Attempts have therefore been made to develop systems for routine use in which other reporter molecules such as biotin are incorporated into the macromolecules. Biotin can be successfully incorporated into both DNA and RNA in vitro (14,15), and the results of hybridization reactions assayed by use of biotin-specific antibodies or avidin-enzyme complexes, to detect the bound probe molecules. At present such systems do not have the sensitivity of labelling methods using radioisotopes. However, more sophisticated detective systems using enzyme amplification or luminescence are being developed which may well increase sensitivity. Until such time however, the use of radioisotopes will continue to be the technique of choice.

7 CONCLUSIONS

The use of radioisotopes has enabled the science of molecular biology to advance over the last 30 years to a stage where molecular approaches are now being used in all branches of biological science. It is now possible not only to understand the role of macromolecules in cellular structure and function, but also to define the defects involved in a wide variety of human diseases. In the case of inherited disorders, radioactively labelled DNA or RNA probes can be used to perform antenatal diagnosis on the foetus as early as 6-8 weeks after conception. It is clear therefore that molecular biology is now moving out of the research laboratory and into clinical, forensic, and analytical use. None of this would have been possible without the use of isotopes, particularly radioisotopes, and it is clear that they will continue to play a major role in molecular biology.

REFERENCES

1.   J.D. Watson and F.H.C. Crick, Nature, 1953, 171, 737.
2.   M. Meselson and F.W. Stahl, Proc. Natl. Acad. Sci. USA, 1958, 44, 671.
3.   M. Meselson and R. Yuan, Nature, 1968, 217, 1110.
4.   S. Cohen, A. Chang, H. Boyer and R. Helling, Proc. Natl. Acad. Sci. USA, 1973, 70, 3240.
5.   J.D. Watson, J. Tooze and D.T. Kurtz in 'Recombinant DNA - A Short Course', W.H. Freeman & Co., New York, 1983.
6.   T. Maniatis, A. Jeffery and D.G. Kleid, Proc. Natl. Acad. Sci. USA, 1975, 72, 1184.
7.   A.M. Maxam and W. Gilbert, Proc. Natl. Acad. Sci. USA, 1977, 74, 560.
8.   D.J. Weatherall, 'The New Genetics and Clinical Practice', Oxford University Press, Oxford, 1985.
9.   P. Gill, A.J. Jeffreys and J.D. Werrett, Nature, 1985, 318, 577.
10. R. Fitts, M. Diamond, C. Hamilton and M. Nevi, Appl. Environ. Microbiol., 46, 1146.
11. J.C. Alwine, D.J. Kemp and G.R. Starke, Proc. Natl. Acad. Sci. USA, 1977, 74, 5350.
12. P.A. Melton, Nucl. Acids Res., 1984, 12, 7035.
13. P.H. O'Farrell, J. Biol. Chem., 1975, 250, 4007.
14. P.R. Langer, A.A. Waldrop and D.C. Ward, Proc. Natl. Acad. Sci. USA, 1981, 78, 6633.
15. A.C. Forster, J.L. McInnes, D.C. Skingle and R.H. Symons, Nucl. Acids Res., 1985, 13, 745.

10
Industrial Applications of Radioisotopes

D.J. Lester

RADIOCHEMISTRY DEPARTMENT, ICI PHYSICS RADIOISOTOPE SERVICES, P.O. BOX I, BILLINGHAM, CLEVELAND, UK

1 INTRODUCTION

Radioisotopes find applications in many diverse industrial fields. Some of the most widely recognised applications by the general public include Energy production, Medical radiotherapy, Isotope production and weapons manufacture. It is impossible to cover all these applications in the space of this short talk so I have limited myself to the areas of industrial application that the ICI Physics and Radioisotope Services is involved with in its daily life. I hope to cover a large variety of our industrial applications to give an overview of some of the fascinating uses industry makes of radioisotopes. The area has been reviewed extensively [1,2] and the reader is referred to these reviews for more detailed information and primary reference sources.

1.1 Types of Radiation. We are all familiar with the three basic types of radioactivity - alpha, beta and gamma. The properties of these types of radioactivity define the use that can be made of isotopes emitting a particular type of radiation. Alpha sources find very limited industrial application due to their extremely low path lengths in air. It is, however, this very property that is exploited in their application as smoke detectors. The relatively large mass of the helium nucleus gives alpha particles considerable momentum and the destructive power in a collision with other matter should not be underestimated. Alpha sources although not of a penetrating nature are highly toxic and not generally used for tracer applications. Most beta sources on the other hand are less toxic than alpha sources, more penetrating and find wide industrial application as radiotracers. Their use is often characterised as non sealed source applications. Two areas of significant importance are in metabolism studies for both new agrochemical and drug development and process flow measurements.

Of much wider application are the sealed gamma sources. The extremely high penetrating properties of this type of radiation is used to examine internal structures and substructures which would otherwise be invisible and as such is a very valuable diagnostic tool. Of comparable penetrating power to gamma radiation are neutrons available either from direct spontaneous fission of an actinide or by the interaction of an alpha particle with a beryllium nucleus. The neutron is uncharged and has a much higher mass than either an electron or proton and can only lose its energy by direct collision with an atomic nucleus. This gives rise to unique applications not possible with the gamma sources.

1.2. Production of Radioisotopes. Before 1930, the only available isotopes were those occurring naturally, Uranium 235, Uranium 238, Thorium 232 and their decay products. The radio-elements are obtained after a lengthy extraction / purification / enrichment process. With the discovery of nuclear fission and subsequent availability of high neutron flux, artificial radioactive materials become available in large numbers. The isotopes are made by a variety of nuclear reactions. Of importance to ourselves as users is the rate at which useful concentrations of isotope can be produced. For example, carbon 14 takes 2-4 years in a high flux reactor to reach useful concentrations. As a result there are only two suppliers world wide.

$$^{14}_{7}N + ^{1}_{0}n \longrightarrow ^{14}_{6}C + ^{1}_{1}P$$

On the other hand bromine 82 can usefully be made from naturally naturally occurring bromine in a few days using a much lower flux. We use the reaction for the production of the relatively short lived bromine 82 usually in an organic form (dibromobenzene) as a basic tracer for many of our process investigation studies.

As a very crude generalisation the production of relatively short isotopes is an easy process capable of being done simply by irradiating the non-radioactive naturally occurring isotope already in the desired molecular form. This is the way most lay people expect us to make radioisotopes. However, in the case of some low molecular weight isotopes like carbon 14, tritium, sulphur 35, this direct irradiation of the desired molecule is not possible. The radioisotope can only be made in a few limited chemical forms. For example, carbon 14 is produced from aluminium nitride as aluminium carbide. Subsequent chemical treatment releases this as carbon dioxide which is adsorbed into barium hydroxide and used universally as a carbon 14 source. All other carbon 14 labelled products are produced from carbon dioxide by chemical synthesis.

The production of the vast variety of diverse carbon 14 compounds becomes an interesting challenge in small scale organic synthesis. Very infrequently does the radiolabel have any influence on the synthesis other than increasing the toxicity of the product with subsequent handling precautions. There is no external radiation risk since even with large sources (5 Curies) the soft beta radiation is completely shielded by the glass containers. Any β particles emitted from the open vessels are stopped by a few decimeters of air.

Carbon 14 labelled products are required extensively by the agrochemical and drug industries who are required by law to carry out a large range of metabolism and degradation studies to demonstrate the fate of a potential new product before it can be safely launched onto the market. The production of suitably radiolabelled tracers with the label in metabolically stable positions presents new challenges since radiochemicals unlike the non-labelled product are not chemically stable over long periods of time. Consequently each product has to be synthesised as required on a one off basis from carbon dioxide. The range of routes available to more readily recognised intermediates is shown in scheme I.

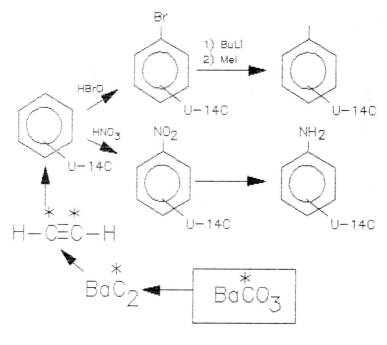

SCHEME

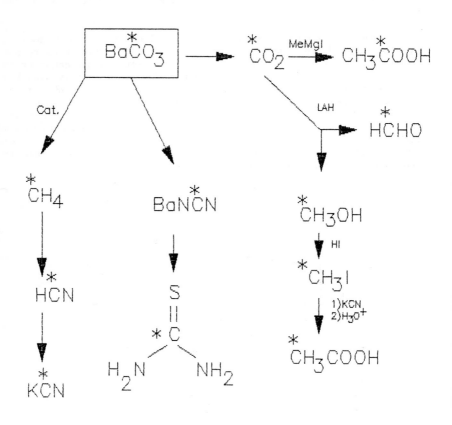

SCHEME 1 (continued)

These building blocks are then used for chemistry often similar to the production route to give specifically labelled products.

1.3. Use of Unsealed Tracers.   There are a large number of industrial examples where chemically simple unsealed tracers are used as an aid to process development / investigation.   These include flow measurement, leak detection and downhole tracers for the oil industries.   The major advantage of using a radiotracer is the ease of detection of the product.   A recent example demonstrating a less obvious use in the commercial environment involved the preparation of a complex carbon 14 labelled fungicide.

White pigments for paint manufacture are stored under water. With time, bacterial growth leads to discolouration of the pigments. Consequently, fungicides have been developed which protect the pigments when used at low concentrations (50 ppm).   The fungicide manufacturer had received several complaints as to this fungicide being inneffective since bacterial growth had been observed on the pigments even when higher concentrations (200 ppm) of the fungicide were used.   It was postulated that the pigments removed the fungicides and so reduced their effectiveness.   A measurement of fungicide on the surface of the wet pigment was required.   Drying of the samples was not desirable since this may have altered the distribution of fungicide between solid and liquid phase.   By preparing a $C-14$ labelled fungicide and also labelling the water with $H-3$ it was possible to design an experiment which conclusively demonstrated the fate of the fungicide and so resolve a potential dispute between fungicide supplier and customer.

2 COMMON INDUSTRIAL APPLICATIONS OF SIMPLE RADIOTRACERS

2.1.   Flow Measurement.   Efficient plant operation requires a complete knowledge of flow rates in either liquid, solid or vapour phases or mixed flows containing two or more phases.   Radiotracer techniques offer methods to:

a)   Measure flow rates more accurately than using existing installed equipment.
b)   Provide data on streams not normally necessary during normal plant operation.
c)   Provide calibration for existing on-line measuring devices.   The methods of measurement fall into two major categories.

2.1.1.   Pulse Velocity Measurement.   In this method the tracer is injected into the process stream as a   slug of activity.   Its appearance between two monitors of known separation   is timed, hence the velocity of the   stream can be calculated.   For use in steel pipes gamma emitting isotopes are necessary with short half lives. They have to be used in a suitable chemical form to ensure they are compatible with the process stream under investigation. Commonly employed isotopes include $Na^{24}CO_3$ ($T_{\frac{1}{2}}$ 15h), NH_4Br^{82} ($T_{\frac{1}{2}}$ 36h), KBr^{82}, [Br-82] methyl bromide, [Br-82] p-dibromobenzene, Ar^{41} ($T_{\frac{1}{2}}$ 1.83h) Kr^{85} ($T_{\frac{1}{2}}$ 10y).

2.1.2. Dilution Measurements. If the radioisotope is admitted at a constant rate into the process stream such that there is complete mixing of the tracer with the stream, then the change in specific count rate of the tracer allows calculation of the process flow rate. This method involves the injection of the radiotracer at constant flow rate and sampling of the stream under investigation at some point down stream where complete mixing will have taken place. The measurement is independant of the cross section of pipe. Errors in the method arise from:

a)   Incomplete mixing of the radiotracer.
b)   Variation in the injection rate.
c)   Measurement of specific counting rate.

The method can be used with any isotope since the radiation is not shielded by process pipework. Tritiated water is often used as a tracer to examine steam flows using this method. It is desirable to have the tracer chemically identical to the process stream since several phases are usually present.

As an extension of flow measurement, radiotracers may be used for residence time studies. These are often important during the design of new plants. The theory governing residence time studies is often complex. The use of a radiotracer in these studies simply increases the ease of analysis of samples for the product under investigation.

2.2. Leak Detection. There are many different techniques widely used for leak detection including search gas, thermal imaging and ultrasonics. Radioactive tracer results are nearly always unequivocal. The tests may be carried out either on-line during normal operation of the plant or off-line during routine maintenance. Although identical in concept these two different situations often allow different sensitivities to be achieved due to the time available for experiment design.

Three major methods are used:

a)   Flow rate measurements before and after the suspected leak. Suitable for large leaks and passing valves.
b)   Changes in residence times.
c)   Direct methods identifying the leaked tracer in a foreign process stream.

By way of example of this type of trouble shooting method let's consider an underground pipe carrying liquid with a leak to the surrounding soil. In a static situation the pipe may be filled with a suitable tracer, eg, Na-24 and pumped up to working pressure. After a suitable time the radioactive tracer is removed from the pipe. The position of the leak can be identified since the radioactive tracer will have leaked into a defined area around the pipe and as such will be detectable from the surface.

If static measurement is not possible then application of pulse velocity measurements may identify the position of the leak since fluid velocity changes rapidly on passing the leak.

The method of static leak testing can be further developed to use sealed sources. By sealing the damaged pipe and continuing to pump a flow towards the leak is established. By introducing a sealed source tracer into the pipe ("pig") it will flow to the position of the leak before stopping, since there will be no fluid flow past the leak. It is then a simple job to identify the leak position by detection of the radiation.

Fluid carryover from steam drums needs monitoring in high pressure steam systems to prevent rapid corrosion. By injecting small ammounts of Na-24 tracer into the steam drum and collecting the steam condensate down stream it is possible to get an accurate measure of carryover, since in the absence of carryover, zero activity should be detected in the condensate.

2.3. Application of Radiotracers to the Motor Car Industry. Radiotracer studies are used in many diverse industries where either accurate flow measurements or extremely low detection limits are necessary. In the motor car industry, the ventilation within the car is difficult to assess. Injecting a radiotracer with suitable radiodetectors dispersed within the seat area allows facile measurement of air distribution and flows which would otherwise be difficult to monitor.

Engine wear can be monitored efficiently using radiotracer techniques. The effects of oils and their additives may be quickly assessed. For example irradiation of a top piston ring in a nuclear reactor produces a whole range of isotopes due to the elements within the steel. After the short lived isotopes have decayed, the sample contains essentially P-32 as a beta emitter and Fe-59 and Co-60, as gamma emitters. Detection of the Fe-59 as it builds up in the oil, gives a fast precise measurement of corrosion occurring. Fractions of milligram loss from the piston ring can readily be detected using this technique.

The technique is used in many other industrial applications to monitor corrosion rates, often by preparing radioactive metal coupons and monitoring their loss into the process stream. The radioisotopes simply allow extremely low detection limits and speeds up what would be a lengthy experiment by conventional analytical techniques.

3 USE OF SEALED SOURCES

The tracer applications mentioned so far result in contamination of the process with the radiotracer. Consequently, due to the strict control of such experiments, unless short lived radioisotopes are used, it can be very time consuming to get the necessary permission to carry out the experiments. Sealed source experiments do not contaminate the process stream under investigation, are often quicker than the tracer experiment and provides complementary information to the tracer technique.

Penetrating radiation (gamma, x-rays or neutrons) is directed at the process vessel or stream under investigation and the transmitted or scattered radiation examined. The sample under investigation does not become radioactive. In the case of neutron sources, theoretically it is possible to form radioactive isotopes within the sample, however, in practice the neutron intensity used is so low that no residual sample activity is observed.

3.1.   Types of Sealed Source.   The chemical form of the radioisotope is of no concern. The important criteria for chosing a particular source depends upon:

a)   Type of radiation.
b)   Radiation energy.
c)   Radiation intensity.
d)   Isotope half life.
e)   Source production cost.

For plant investigations, point sources are desirable. The isotope pellet is contained within a stainless steel capsule. Most useful gamma sources either have low (<200 KeV) or high (>0.5 MeV) energies with few sources available of intermediate energy.

Neutron sources are interesting since few useful isotopes emit neutrons. The most commonly used single isotope neutron source is californium 252 with a half life of 2.65y. The isotope produces soft neutrons with a low gamma emission and consequently is relatively safe to handle. Its high cost and shortish half life are its major disadvantages.

A more common neutron source consists of an alpha source mixed with beryllium. Interaction of alpha particles with beryllium produces neutrons of low energy spread, a desirable characteristic.

The sources are referred to as "alpha-n" sources.

$$_{4}^{9}Be + _{2}^{4}He \longrightarrow _{6}^{12}C + _{0}^{1}n$$

The choice of a particular source for a given application depends upon many factors. For continuous on-line measurement or control softer, low energy sources are desirable whilst non-invasive trouble shooting methods often use much more energetic sources with long ranges. Measurements are usually made on the attenuation or backscatter of the radiation beam. It follows that careful selection of the beam intensity is necessary for meaningful accurate measurements.

3.2. Typical Plant Measurements Using Radioactive Sources

3.2.1. Gamma Absorption Techniques. This is a simple concept with several applications. Carefully chosen penetrating radiation is passed through the process vessel or pipe containing the process fluid. The emerging radiation is detected using either a sodium iodide scintillation counter or geiger muller tube and attenuation of the beam calculated. Since the degree of attenuation is related to both the thickness of the vessel and the density of the contents, known information on either element allows measurement of the variable. Information on material thickness, contents density or changes in density profile can be obtained easily. The most commonly used isotopes are Co-60, Cs-137 and Am-241. The physical set up is simple but no single measurement gives unambiguous results. It is usual to map the profile of the pipe or vessel at several points to get a clear unambiguous diagnosis. Important applications include thickness measurement without plant shutdown, quality control of metal sheet thickness, corrosion measurement, checking for voids in concrete and assessing the depths of deposits on the inner walls of pipes and vessels. Detection of pipe deposits is important in maintaining high plant efficiency. Small deposits on heat exchangers give rise to very poor heat transfer properties, whilst build-up in pipe lines increases pressure drop and ultimately results in blockage. In gas lines, accumulation of liquid may result in slugs which can cause extensive damage down stream due to hammer action.

Corrosion measurement is obviously important. Unsealed applications already discussed have the most sensitivity. Gamma adsorption techniques find application in routine examination of heat exchanger bundles and application to inaccessible areas, for example irregular regions around pipe welds.

In cases where the thickness is known, or is not a variable then changes in radiation beam intensity relate to density changes in the process stream. Installed gauges allow continuous monitoring of process density.

Performance of catalyst beds and packed distillation columns depends upon even packing, ie, packing density. Measurement of the catalyst bed profile or distillation column profile to adsorbed gamma rays by scanning the length of the column with source and detector through the width of the column allows irregularities to be identified and process inefficiencies hopefully improved. It should be realised that in many cases we are not looking for small irregularities. When processes show reduced efficiency it often relates to major failure inside columns resulting from completely collapsed supports.

Liquid level measurement either as a continuous on-line measurement or for diagnostic purposes looks at the step change in density at the liquid/vapour interface. By using a wide diverging beam of radiation covering the height of the vessel and an array of detectors up the vessel, a continuous non-invasive nucleonic level gauge is easily installed on the plant.

3.2.2. Radiography. An important application which in fact represents a field in its own right is radiography. Here attenuation of the radiation beam is detected on a photographic film thus building up a visual picture of the internal structure of the vessel or homogeneity of the construction material. It finds wide application in non-destructive testing of pipeline welds.

Photographic film is relatively insensitive to radiation compared to electronic detection. To obtain reasonably short exposure times (a few minutes) it is necessary to employ large high energy sources with the resulting health and safety hazards. It is necessary to ensure that the film is in close contact with the object to ensure good definition. Looking inside pipelines in this way often reveals "foreign" objects causing flow problems, eg, spanners left behind by shutdown engineers.

3.2.3. Mass Flow on Conveyor Belts. Extension of the idea of density measurement in the liquid and gas phases to the solid state identifies a common measurement made by the mining, steel making and construction industries.

Continuous flow of irregular shaped coal on conveyor belts when scanned continously with a gamma radiation beam presents a detector with a changing signal which is related to the mass of material between the detector and source. It is therefore possible to use radioactivity to continuously measure the mass on the belt with no need for any physical contact between the belt and weight sensor. This method presents attractive savings in maintenance over conventional methods. The measurements are unaffected by belt tension and offer a considerable improvement over conventional methods.

3.2.4. Radiation Scattering Techniques. The methods discussed so far have been made simply by interposing the process vessel between source and detector. This has resulted in simple measurements with results that are easily interpreted. Sometimes it is not physically possible to straddle the vessel with source and detector or the vessel is so large that a large source would be necessary to allow the radiation to completely penetrate the vessel. In these cases the detector, suitably shielded is placed alongside the source and the backscattered radiation examined. The methods will give similar results to transmission methods except that the data is much more difficult to interpret. The set up is critically dependant upon its geometry. Level measurement is usually simple, however, using this technique.

Since part of the backscattered radiation has resulted from interactions of radiation with the inner electron orbitals, it is characteristic of the elements present and may be used as a qualitative analysis of metals present. Portable x-ray fluorescent analysers are employed when such on-line measurement are necessary.

3.2.5. Neutron Techniques. The recent advances in the production of neutron sources and their detection has led to a fast growing use of neutron techniques. Neutron techniques offer completely different absorption characteristics by more than alpha or gamma rays and can often allow data to be obtained that would not be possible using gamma ray adsorption techniques.

Fast neutrons have no charge but a large mass and momentum. They are not affected by electron collision, consequently have a large range in matter.

Fast neutrons (0.5-11 MeV) undergo both elastic and inelastic collisons with other atomic nuclei resulting in energy loss and slowing down of the neutrons. This process is accompanied by emission of gamma radiation. Slowing down to energies of about 1 eV produces epithermal neutrons. Further collisions result in the production of thermal neutrons of energies 0.25 eV. The slowing down of epithermal and thermal neutrons depends upon the hydrogen concentration in the matter with which the neutrons interact giving rise to the completely different adsorption characteristics than shown by gamma radiation.

The three main important features of neutron interaction may be summarised:

a) The "moderation" (ie, slowing down) of fast neutrons by a medium to produce thermal neutrons is primarily determined by the hydrogen concentration in that medium to a first approximate independantly of its chemical form.

b)   Some elements are extremely effective absorbers of slow
     neutrons. Low levels can significantly reduce neutron flux.
c)   The adsorption of a neutron by a nucleus usually produces a
     radioactive isotope which upon decay can be fingerprinted by its
     characteristic radiation spectrum.

The basic method of using neutron adsorption techniques or
backscatter techniques is similar to the corresponding gamma ray
technique. A suitable detector is a helium-3 proportional counter.

3.2.6.   Moisture Meter.   The properties of neutrons lend
themselves to the easy detection of water and have found application
in the in situ measurement of water in soil. The major user of this
technique is the steel industry, making measurements on the moisture
content of coke and sinter mix charged to the blast furnace. Neutron
gauges are used to control the water/cement ratio in ready mix
concrete.

3.2.7. Level and Interface Measurement. Neutron techniques are
useful in determining level interfaces between process fluids of
similar densities inside process vessels provided the two phases have
different H-concentrations, particularly if one of the layers is
water.   Neutron backscatter techniques can be used quickly and
effectively although their use, particularly as installed gauges is
not as common as the gamma ray devices. The reasons for this are
best understood by looking at some of the advantages and
disadvantages of the method.

Firstly, the advantages:

a)   Non-invasive technique.
b)   Independent of vessel diameter.
c)   Measurements made from one side of vessel.
d)   Measurements made by one operator.
e)   Method distinguishes between materials having different neutron
     moderation properties.

and the disadvantages:

a)   Difficult to apply to thick walled vessels.
b)   Lagging makes the technique insensitive.
c)   The result is not representative of the interface across the
     whole vessel.
d)   Only useful for hydrogenous materials.
e)   Equipment for the technique is expensive.

3.2.8.   Neutron Activation Technique.   This again is a large
field in its own right.   Interaction of matter with high neutron
fluxes (10^{12} slow neutrons cm^{-2} s^{-1}) found within the core of a
nuclear reactor produce a whole range of radioactive nuclei.   By

measurement of the X-ray energies of these decays an extremely
sensitive analytical technique is available offering rapid analysis
of a range of elements in the same sample.

4   SUMMARY

Techniques involving radioactivity offer unique ways of process
examination. An overview of the application of these techniques has
been presented. It is only after several years of development and
experience that the techniques offer sound advice for the improvement
of process plant efficiency.   ICI have learned and developed these
techniques over many years and now offer them as a world wide service
to industry.

5   REFERENCES

1   J A Heslop, Chemical Society, Specialist Periodical Reports,
     Radiochemistry 1976 volume 3 "Industrial Applications of
     Radioisotopes".

2   "Radioisotope Techniques for Problem Solving in Industrial
     Process Plants", Editor; J S Charlton, Leonard Hill, 1986.

Index